ANIMAL C

DATE DUE

MR 17 '95		
DE 9 '97		
MR 31 99		
MY 11 00		
28 05		

DEMCO 38-296

The INTRODUCTION TO BIOTECHNIQUES series

Editors:

J.M. Graham MIC Medical Ltd., Merseyside Innovation Centre, 131 Mount Pleasant, Liverpool L3 5TF

D. Billington School of Biomolecular Sciences, Liverpool John Moores University, Byrom Street, Liverpool L3 3AF

CENTRIFUGATION

RADIOISOTOPES

LIGHT MICROSCOPY

ANIMAL CELL CULTURE

Forthcoming titles

GEL ELECTROPHORESIS: PROTEINS

GENE TECHNOLOGY

PLANT CELL CULTURE

PCR

ANIMAL CELL CULTURE

Sara J. Morgan
Leukaemia Research Fund, 43 Great Ormond Street,
London WC1N 3JJ, UK

David C. Darling
Medical Molecular Biology Unit, University College and
Middlesex School of Medicine, The Windeyer Building,
46 Cleveland Street, London W1P 6DB, UK

In association with the Biochemical Society

ιay be reproduced or transmit-
permission.

993 by

OX1 1SJ.

A CIP catalogue record for this book is available from the British Library.

ISBN 1 872748 16 3

Typeset by Westwater Phototypesetting, Frome, UK
Printed by The Alden Press Ltd, Oxford, UK

Preface

Animal Cell Culture is intended as an introductory book for those who are new to cell culture, whether undergraduates, graduates, technicians or post-doctoral scientists.

As researchers who have both spent much of our time at the bench performing cell culture, we recognize the need for an inexpensive, introductory text for the inexperienced cell culture worker. Excellent, comprehensive works are already available in this field, but the sheer number and variety of current techniques in animal cell culture has resulted in long and expensive texts. This book is not meant to be a comprehensive list of every available protocol; instead, it aims to explain why and how the basic techniques are used, and their applications in modern cell and molecular biology.

The initial chapters describe the basic methodology needed for the culture of cells, including aseptic technique, media preparation, and guidelines for the correct use of equipment. The basics of performing cell culture on both established and primary cell lines are described, together with the essentials of spotting and dealing with contamination. The storing of cells by cryopreservation completes the suite of basic essential techniques. Armed with the necessary background, these techniques can be applied to the cloning of cells, the separation of lymphocytes from various sources and the use of these in cell fusion, cytotoxicity assays and the immortalization of human B-lymphocytes. Finally a brief overview (with selected techniques) is given of the expanding field of animal cell transfection.

It is hoped that this book may act as a stimulus for further, more detailed, reading. Theoretical considerations have been included where necessary for a better understanding of the techniques described. Much of what is contained in this book is the result of the accumulated knowledge and experience of countless researchers who have refined techniques to the point where they can give reliable results even in the hands of inexperienced workers. To paraphrase Newton: a modern scientist sees into the future only by standing on the shoulders of giants.

S.J. Morgan
D.C. Darling

Acknowledgments

We are grateful to Dr Amata Hornbruch, Department of Anatomy and Cell Biology, UMDS, Guy's Hospital, London for the original details of chicken embryo fibroblast isolation and Esther Bell, Department of Anatomy and Developmental Biology, University College and Middlesex School of Medicine, London.

We are grateful to Dr Guy Whitley, Department of Cellular and Molecular Sciences, St. Georges Hospital Medical School, London, for his help in providing information on the use of human tissue for the isolation of primary cells.

Many thanks to Jenny Tooze, Division of Haematology, St. George's Hospital Medical School, London, for providing rosetted T-cells.

We are grateful to David Billington for the method used for isolating hepatocytes.

Contents

PART 2: TECHNIQUES AND APPLICATIONS

11. Cell Fusion 109

12. Cytotoxicity Assays 121

13. Epstein–Barr Virus Immortalization 129

Abbreviations

AET	2-aminoethylisothiouronium bromide
BME	Eagle's basal medium
CAT	chloramphenicol acetyltransferase
CMFH	calcium- and magnesium-free Hank's balanced salt solution
ConA	Concanavalin A
CTL	cytotoxic T-lymphocyte
DMEM	Dulbecco's modified Eagle's medium
DMSO	dimethylsulfoxide
EBSS	Earle's balanced salt solution
EBV	Epstein–Barr virus
ECGS	endothelial cell growth supplement
EDTA	ethylenediaminetetraacetic acid
EGF	epidermal growth factor
ELISA	enzyme-linked immunosorbent assay
FCS	fetal calf serum
FGF	fibroblast growth factor
FSH	follicle-stimulating hormone
GM-CSF	granulocyte/macrophage colony-stimulating factor
HAT	hypoxanthine, aminopterin and thymidine
HBS	Hepes buffered saline
HBSS	Hank's balanced salt solution
HCM	'half-conditioned' medium
Hepes	N-2-hydroxyethylpiperazine-N'-2-ethanesulfonic acid
HGPRT	hypoxanthine guanine phosphoribosyl transferase
IGF-II	insulin-like growth factor II
IL-2	interleukin-2
IL-3	interleukin-3
IMS	industrial methylated spirit
LAK	lymphokine-activated killer
Mops	3-N-morpholinopropane sulfonic acid
NCS	newborn calf serum
NGF	nerve growth factor
NK	natural killer
PBS	phosphate-buffered saline
PHA	phytohemagglutinin
TPA	12-O-tetradecanoylphorbol 13-acetate

1 Introduction

1.1 Why culture cells?

What is the purpose of cell culture and how is it performed? Cell culture (or 'tissue culture' as it is also known) has its origins in the 19th century when people began to examine in some detail the tissues and organs of the body in glass vessels. The term *in vitro* literally means 'in glass', although today most cell culture is performed in or on plastic. Most cell culture media currently in use are based upon original work on the formulation of solutions that allowed cells from tissues to survive for short periods outside the host body. With the modifications that resulted in more sophisticated media it was possible to grow tumor cells from malignant tissue samples from both humans and animals [1,2]. Cell culture has now arrived at a position where often the objective is to reproduce the *in vivo* conditions to allow 'normal' (non-tumor) cells to be grown in culture. Cell culture has thus progressed from the examination of isolated cells as they die *in vitro*, to the continuous (or long-term) culture of cells that are in many respects identical to those found in the whole organism.

Cells from a wide range of different source tissues and organisms can now be grown in the laboratory, but why is this done? Originally the major purpose was to study the cells themselves, how they grow, what they require for growth, and how and when they will stop growing. These types of study are still very much a research interest today, for example in the areas of the cell cycle, the control of tumor cell growth and the modulation of gene expression.

Another major area of interest is in developmental biology. Attempts to explain how the multitude of cells present in the mature organism are derived from the single cell at fertilization have prompted many to look for experimental models. Cell culture is well suited as a model for the study of development and differentiation, and cell lines that retain the ability to differentiate *in vitro* are the subject of intensive research [3]. Finally, there are some types of work that simply cannot be done without cell culture, for instance, work with transgenic animals that results in mature organisms expressing novel genes is absolutely reliant on cell culture techniques for the insertion of the foreign gene into the recipient cells [4,5]. Also, cell fusion technology and cytotoxicity testing are both cell culture techniques that have to some extent been designed to replace the *in vivo* methodology.

1.2 Aspects of cell culture

This book presents the fundamentals required for cell culture; the novice is introduced to the equipment required (Chapter 2) and the ideal environment in which it should be placed. Familiarity with the practicalities of aseptic technique is vital – the manipulative skills required for efficient contamination-free cell culture are illustrated in Chapter 3. The methods for preserving the sterility of cell cultures during manipulation can become automatic and will prevent the waste of both time and resources. We hope that the illustrations will allow those practices most likely to cause problems to be kept to a minimum.

The choice and preparation of cell culture media can be very confusing. The selection of media type and choice of starting points for the preparation of complete media are examined, the preparation of cell culture media is described, and details of the most often used supplements are given in Chapter 4. The chapter covering the culturing of cell lines addresses the everyday procedures in maintaining healthy cell cultures, the subculturing of cells and the construction of growth curves. Primary culture techniques merit a book to themselves: they are as varied as the sources from which they are derived. They require skills in dissection and cell manipulation beyond the scope of this book; nevertheless, the basic techniques are explained to prepare and culture primary cells from a number of sources.

Every cell culture worker has had contamination problems, but it happens less often with experience. It is necessary to identify the contamination and the possible source before problems can be rectified. To help with this, Chapter 7 introduces the most common contaminants and suggests what (if anything) can be done about it. The ultimate safeguard against the loss of valuable cultures by contamination is cryopreservation (covered in Chapter 8); this is also a basic culturing technique that will minimize waste of both time and resources caused by growing cells which are not required immediately.

The culturing of cells is usually performed with a specific objective in mind, and therefore a number of more advanced and applied techniques are included. Lymphocyte Separation and Guidelines for Establishing Lymphocyte Lines (Chapter 10) allows one to isolate and separate different lymphocyte subpopulations from a range of different sources, and advises upon the best source material to suit particular needs. These lymphocytes can be used for the propagation of antigen-dependent T-cell lines, cell fusion (Chapter 11), cytotoxicity studies (Chapter 12) and the Epstein–Barr virus (EBV) immortalization (Chapter 13) of human B-lymphocytes. Cloning Techniques (Chapter 9) introduces the different methods available for the isolation of single daughter cells from cell cultures, and discusses the suitability of each different method. This important technique is an integral part of many of the advanced and applied methods described in Part 2 of the book, for example, the cloning of a hybrid cell line (Chapter 11).

Finally, transfection (Chapter 14), a technique that owes a lot to other branches of biology but which is intrinsically dependent upon cell culture is reviewed. The reader is introduced to the more common molecular biological techniques and terminology used today, and their relationship to the transfection of animal cells is detailed.

The techniques described in this book are all in common use and therefore it follows that there are many possible modifications. We encourage experimentation, as it is possible that, for your own cell system, improvements can be made.

Probably the least interesting – but definitely the most important – aspect of cell culture is safety; we urge you to read and act upon the advice that follows. There will also be experienced workers around you: ask their advice, their experience is invaluable and will be of great help to you in the future. Local safety officers can answer your queries and worries about the safety aspects of cell culture, and they should be consulted before problems arise.

1.3 Safety considerations

It is important to be aware of the potential dangers that this type of work involves. Safety in the laboratory must be the prime consideration in animal cell culture. Human cell lines present obvious dangers, as they may contain pathogenic organisms which can be shed into the medium. Fresh human tissue such as lymphocytes can, for example, contain infectious agents such as HIV and/or hepatitis B virus.

Infectious agents, when released into tissue culture medium from cells or cell lines can be dangerous, as *aerosols* (fine mists or sprays resulting from agitating or disturbing a liquid) may result from routine manipulation. These aerosols can infect via contact with abrasions or mucous membranes. Obviously agents that use humans as hosts are especially dangerous, as they may be capable of self-replication in a new host.

Fresh (primary) human tissue should be treated with caution: gloves and masks should always be worn when working with these samples. Where possible, the infection status of the donor should be assessed prior to work commencing. A recirculating Class II tissue culture hood is the *minimum* containment level for this type of material (see Chapter 2 for a review of hoods available and recommendations on their use). Human tissue culture cell lines must also be manipulated in recirculating Class II tissue culture hoods. The origin of these cells suggests a potential capacity for transmitting infectious agents.

Non-human cells or cell lines present a lesser danger, although it should be remembered that some primate lines can harbor viruses. In general, animal cells or cell lines (particularly rodent) present few problems to the careful worker, as it is unlikely that contaminating cells would escape host immunologic defenses.

The most potentially dangerous cell types are those whose role is to facilitate the study of human pathogenic organisms, for example HIV or the pathogenic parasite *Trypanosoma cruzi*, the causative agent of South American sleeping sickness. These are absolutely dependent on live cells for at least some part of their lifecycle, and tissue culture is the only method by which they can be propagated.

Often overlooked are the toxic compounds that are routinely used in tissue culture. Compounds such as 12-*O*-tetradecanoylphorbol 13-acetate (TPA) or dimethylsulfoxide (DMSO) should be treated with caution. Accidental exposure to these types of agent is usually a 'one-off' problem resulting in a small single dose. The precautions that should be taken when using these compounds can usually be found in the relevant catalogue.

1.4 Basic precautions for cell culture work

- The correct containment category should always be used; if unsure, check with either your supervisor or local safety officer.
- A protective laboratory coat should always be worn.
- Gloves should be used for all culture work; masks are also advised.
- Beware of contaminated sharps, e.g. syringe needles.
- All liquid and solid waste should be disposed of according to established procedures at your work place.

A waste beaker containing a quantity of 10% hypochlorite solution should be placed in the hood when doing culture work. All liquids should be disposed of in hypochlorite: pipets should be neutralized by taking up the solution as a 'hypochlorite rinse', and flasks and tubes should also be rinsed. Liquid waste after treatment with hypochlorite can be poured away with copious amounts of water (some regulations specify that liquid waste should be autoclaved before disposal).

Much of the work involving cell culture is subject to local and national regulations. Consultation with the relevant authorities – initially the local biological safety officer or equivalent – is advised if there are doubts about safety procedures in specific circumstances.

References

1. Gey, G.O., Coffman, W.D. and Kubiek, M.T. (1952) *Cancer Res.*, **12**, 264.

2. Moore, G.E., Sandberg, A.A. and Ulrich, K. (1966) *J. Natl. Cancer Inst.*, **36**, 405.

3. Sachs, L. (1987) *Cancer Res.*, **47**, 1981.

4. Bradley, A., Evans M., Kaufman M.H. and Robertson, E. (1984) *Nature*, **309**, 255.

5. Capecchi, M.R. (1989) *Science*, **224**, 1288.

2 Equipment and General Practice

This chapter addresses basic guidelines for the use of essential major items of cell culture equipment, such as incubators and tissue culture hoods. This background knowledge can help prevent problems with culture work, including such things as contamination and poor cell growth.

2.1 The cell culture laboratory

An ideal cell culture laboratory is described below; it is recognized, however, that few laboratories will be lucky enough to have all the desirable features. The dimensions and situation of the designated area are both important factors which are often overlooked. Cell culture equipment is often found in a corner of an existing laboratory, with little thought given to the problems that this situation may cause the cell culture worker.

The room should be of adequate size to accommodate the necessary equipment, and isolated from other rooms by at least one – and preferably two – doors (this is sometimes required under local regulations governing tissue culture). These doors should be situated far enough apart to allow the intervening area to be used as a 'changing room' where normal laboratory coats can be changed for those used only in the tissue culture room. Windows in the doors, to allow observation of personnel within and outside the laboratory, can help prevent both doors opening simultaneously and minimizes the entry of possible contaminants into the laboratory. The atmosphere of the tissue culture laboratory should be at a slightly higher pressure than that of the outside – i.e. positive pressure – to prevent inflow of air.

The floor should be easy to clean, non-porous and light in color, to allow the detection of spillage. It is best to have windows (though these should not be opened), because adequate lighting is important: cell suspensions are easier to examine with natural back light.

When all the equipment is switched on and the sun is shining, it will become extremely warm inside protective clothing, and therefore air conditioning is invaluable. Remember, though, that the air-conditioning system must also filter the air, and the positioning of the unit should be carefully considered to prevent interference with the air flow of the hoods.

2.2 Autoclaves and hot-air ovens

These two items should not be situated in the cell culture room, but in a conveniently close location. They are used to eliminate potential 'contaminants' (e.g. bacteria) from cell culture items such as bottles, other glassware, or water for media. Fungal spores are very resistant to assault and the routine conditions of the autoclave/hot-air oven must be designed to eliminate them. A typical autoclave is shown in *Figure 2.1a*.

2.2.1 Principles of operation and safety considerations

Autoclaves are many and varied: the more complex types can perform a number of different programs (or cycles), but the basic principles remain the same. Once the autoclave is sealed, steam is introduced into the system, either directly from an external source or by boiling water from an internal source. Initially the steam is used to purge the vessel of air; once this is achieved the temperature (and thus the pressure) increases to the preset level, normally 121°C and 15 lb/in^2 (103.5 kPa), and is maintained for 15 min. After this period the steam supply is terminated, the pressure drops and the autoclave cools. Bacteria, fungi, their spores, viruses and mycoplasma will not survive these conditions, as the use of 'wet heat' is an effective method for denaturing proteins.

The use of high temperatures and pressures means that autoclaves are potentially dangerous pieces of equipment, and it is important to be familiar with the full manufacturer's instructions before use.

Hot-air ovens are relatively simple pieces of equipment that heat the air inside to a given temperature (180°C), and maintain that temperature for the required time (1 h for sterilizing conditions). Ovens, though not actually essential to cell culture work, are straightforward and useful for some items. Heat killing is less efficient in dry conditions and therefore a higher temperature and longer time are required than when using an autoclave.

2.2.2 What can be put in autoclaves/hot-air ovens?

Hot-air ovens are useful for some glassware, e.g. pipets, beakers and also Duran/Pyrex medium bottles that have 180°C-resistant caps (usually red in color). Some of the older autoclaves without a vacuum/drying cycle can leave moisture in the glassware, which can affect cell culture; this can be avoided in dry sterilization. Other items such as plastic micropipet tips, water, phosphate-buffered saline (PBS) and 50 ml polypropylene centrifuge tubes should only be sterilized under autoclave conditions. Some liquids cannot be autoclaved, e.g. most growth media, glutamine and trypsin, as their components can be destroyed or denatured under the extreme conditions. Plastic items made from polystyrene (i.e. most plastic cell culture ware apart from 50 ml 'Falcon' tubes available from Becton-Dickinson) will not survive autoclave conditions. Naturally, those items that can survive hot-air oven conditions can also be put in the autoclave.

Bottles to be autoclaved should have their caps slightly loosened; other items should not be sealed tightly or sterilization may be ineffective. It is possible to use sterilization bags (available from hospital equipment suppliers, see Appendix B) in which to autoclave more awkward shapes such as large-volume filters (see Chapter 4). Metal pipet cans are available which are suitable for both the hot-air oven and the autoclave.

2.2.3 Testing for a successful autoclave run

Modern autoclaves have sensors within the pressure vessel which record internal conditions throughout the sterilization process; these are often probes located in the middle of a load, and a printout is available to check that adequate conditions were achieved (*Figure 2.1b*). In the absence of such sensors, alternative, less accurate methods can be used:

(1) Autoclave tape, which is initially plain, but on exposure to high temperature produces dark stripes across the tape. This is not the most reliable method of monitoring the sterilization and is better used simply as a marker to indicate those items that have been through a sterilization process.
(2) Sterilization tubes, which change color after exposure to appropriate autoclave conditions. They can be placed inside bottles to be autoclaved and monitor conditions within the bottles.
(3) Autoclave strips: these turn blue progressively along their length, rather like a thermometer, and show safe sterilization when the blue reaches the indicated 'safe' mark.

Modern autoclaves have a drying cycle and do not allow the pressure vessel to be opened until it has dropped to room temperature and pressure. With older models it is possible to open them before they are cool, but this can cause a rapid change in temperature and pressure, resulting in the breakage of glassware and possible contamination, as non-sterile air is suddenly drawn into the autoclave.

Once the autoclave has been opened, any caps which were loosened prior to the sterilization should be closed.

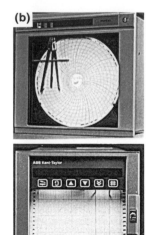

FIGURE 2.1: Autoclaves. **(a)** An example of an autoclave (courtesy of Astell Scientific). **(b)** Many autoclaves have a chart recorder to check that the autoclave is performing to specification (courtesy of Astell Scientific).

2.3 Tissue culture hoods

There are two principles considered in hood design:

(1) protecting the tissue culture from the operator (ie. a sterile environment),
(2) protecting the operator from the tissue culture (in situations of potential infection risk).

Each type of hood addresses these principles to varying degrees. A brief introduction to the various hoods available is given below.

2.3.1 Different types and principles of operation

Tissue culture hoods can be grouped as followed:

(1) laminar flow (*Figure 2.2a*)
(2) class I
(3) class II (*Figure 2.2b*)
(4) class III.

Figure 2.3 illustrates air flow in Class II and laminar flow hoods. Laminar flow hoods offer no protection to the tissue culture worker: the cabinet is open and a stream of filtered and sterile air flows from the back, over the workspace towards the operator. Obviously any cell culture that harbors a potential risk should not be used in these hoods, and it is recommended that only media-making and sterile filtration activities are performed in these hoods. The open design of the hood makes it useful for manipulations of this type.

Class I hoods give good protection to the operator and, to a lesser degree, the cell culture. Air is drawn from the open front (past the operator) over the cell culture, and out through the top of the hood. These hoods are found within specially designed sterile work areas (i.e. sterile air is sucked into the hood), where users wear special protective clothing.

Class II hoods offer protection to both the operator and the cell culture. Filtered air is drawn in through the top of the hood, down over the tissue culture and through the bottom of the work area. In addition, air is drawn from the half-open front past the operator and down through the grill in front of the work area. In this way the cell culture is protected in a stream of sterile air and the operator is protected from contamination by the inflow of air into the base of the work area. The Class II hood is the most common type found in a tissue culture laboratory.

Class III hoods are used for work with highly pathogenic organisms. In these, the worker is screened from the work by a full physical barrier. This is normally achieved by the replacement of the open front with glass or perspex, with a pair of heavy-duty protective gloves attached, through which the work is accessed. The use of a Class III hood requires know-

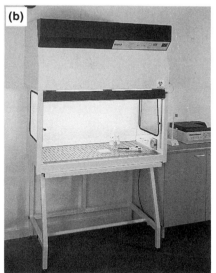

FIGURE 2.2: Tissue culture hoods/cabinets (courtesy of ICN Biomedicals). *(a)* A laminar flow (horizontal) hood, which gives no operator protection but good sterile protection. *(b)* A safety (Class II) cabinet which gives good operator protection and good sterile protection for the cell culture. Some tissue culture hoods now look similar to a Class II safety hood but are in fact vertical laminar flow hoods; care must be taken to assess the suitability of the hood for the work.

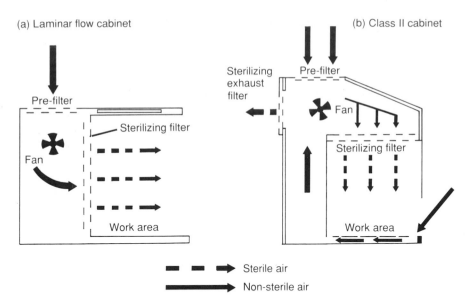

FIGURE 2.3: Air flow diagrams. (a) Horizontal laminar flow cabinet and (b) a Class II recirculating cabinet.

ledge of special safety considerations – these should be raised with the relevant local biological safety officer.

2.3.2 Setting up and cleaning out

It is wise to spend some time understanding the hood to be used: note the air inlets and the exhaust ports, and any dials offering information on the air flow and running efficiency. The inlet and exhaust areas should not be restricted, otherwise the hood will not operate at optimum efficiency. Where a hood has a cover available to protect the working area, it is good practice to use it.

To start, the protective cover should be removed and the hood switched on; it should be left to run for a few minutes to allow the air flow to stabilize. Some hoods have sensors to assess the input and output of air and pressure on the filters: the data, which are displayed on a meter, can indicate wear on the pumps and the integrity of the filters. The interpretation of these measurements depends upon the type of hood: help with this can be found in the operator's manual or by contacting the manufacturer's technical department. If you think that the hood is not working properly, do not use it and obtain advice.

Once stabilized the hood can be swabbed with liberal amounts of 70% (v/v) industrial methylated spirit (IMS), this is most easily done with a domestic houseplant spray and absorbent tissue. All internal surfaces should be treated, including the internal surface of the transparent front panel, but do not spray the filter area, where damage may be caused to the filter unit. When work has finished the hood should be completely cleared – do not leave anything inside – and swabbed again with 70% IMS. The motor can then be turned off and the cover replaced.

2.3.3 Safety aspects, maintenance and checking

Hoods should be serviced at regular intervals (usually every 6 months) as part of a service contract. This is often arranged by an institute to include all hoods (usually the cheapest option). It is important to check that the hood you use has been serviced recently.

Although servicing will ensure that the equipment will perform to specifications, this will not prevent contamination problems that arise through bad practice. A common source of contamination is through improper cleaning of spillages: they should be wiped up and swabbed with 70% IMS. If the base can be removed check the underside and clean if necessary. Any removed parts should be cleaned before replacement and then swabbed once again with 70% IMS.

2.4 Incubators

2.4.1 Incubation conditions

For optimal growth most animal cell types require a temperature of 37°C in medium at pH 7.0–7.2. The nature of the by-products of cellular metabolism tends to make the medium more acidic than is desirable, and therefore a buffering system is used. Most commonly this is a bicarbonate/CO_2 system requiring a supply of CO_2 to the incubator and bicarbonate in the medium. Additionally, a highly humidified atmosphere helps prevent evaporation of the medium. However, some laboratories prefer to maintain 'dry', non-gassed incubators, using alternative buffering systems such as morpholinopropane sulfonic acid (Mops) or N-2-hydroxyethylpiperazine-N'-2-ethanesulfonic acid (Hepes) (see Chapter 4). CO_2 incubators control both the internal temperature and the gas mixture in the chamber: these are normally set to a mixture of 5% CO_2/95% air at 37°C, with a water reservoir giving high humidity. The 5% CO_2 must be carefully controlled, as it governs the stability of the buffering system. The bicarbonate added to the medium (as sodium hydrogen carbonate) effectively 'mops up' the acidic ions (H^+, from cellular metabolism), forming carbonic acid which is in equilibrium with the water and dissolved CO_2; this maintains a medium buffered at a pH of 7.0–7.2. The high humidity within the incubator allows the culture medium to be exposed to the CO_2 atmosphere while minimizing evaporation. *Figure 2.4* shows a typical CO_2 incubator.

Dry non-gassed incubators are essentially nothing more than large boxes in which the internal temperature can be set. Cultures are grown in a

FIGURE 2.4: A typical CO₂ gassed incubator (courtesy of LEEC Ltd).

closed system, i.e. the caps of flasks are screwed down. This method of culture is popular with some workers, who feel that water in CO_2 incubators can be a source of contamination. However, the low humidity in the incubator makes it difficult to culture cells in plates, as evaporation of the medium can harm the cultures. In addition, when culture flasks with closed caps are initially placed in the incubator the warming effect will cause the internal pressure to rise, and thus after a period of time the flask will need to be vented.

2.4.2 Setting up a gassed incubator

Each incubator will come with its own particular instructions, which should be studied in detail. However, a brief overview will be given of the most likely sequence of events. The first task is to wash the incubator interior with a mild detergent solution, being careful not to scratch the surfaces, and then swab with 70% IMS. Distilled water should be placed in a plastic tray in the bottom of the incubator; the incubator is then switched on with the temperature set to 37°C, but the CO_2 supply is left switched off at this stage. A thermometer should be placed in the incubator, most usefully on the glass interior door, to confirm the temperature readout of the incubator. A water jacket, if the incubator has one, is an effective insulating device to maintain temperature, and should be filled to the recommended level. The incubator should now be left to stabilize for a few days. The temperature setting will then probably need 'fine tuning' and each time this is done the incubator should be allowed to settle for at least 24 h. If the internal door has a build-up of condensation, the door heater should be turned up: there is usually a small thermostat in the door itself.

Gas cylinders are very heavy, often about 152 cm tall and highly pressurized, therefore they must be firmly secured for safety reasons. A standard pressure regulator needs to be attached to the head of the cylinder: the two gauges show internal pressure of the cylinder and the pressure of the gas leaving the cylinder (*Figure 2.5*). The gas supply from liquid/gas CO_2 cylinders must be connected to the incubator with pressure-resistant tubing.

CO_2 cylinders containing both liquid and gaseous CO_2 will give a high pressure reading until the last of the liquid is exhausted, after which the pressure drops rapidly. It should be recognized, therefore, that the pressure sensor in the cylinder regulator will only indicate that the cylinder is not empty and not how much supply remains.

Most incubators are claimed to be factory-calibrated for 5% CO_2 in air in high humidity, so once the gas supply is correctly connected and the required CO_2 atmosphere set on the incubator, the gas supply can be opened at the cylinder head (the CO_2 supply switch on the incubator should still be off). The supply of gas to the incubator should be at the recommended pressure as indicated (either in the manual or near the CO_2 indicator dial); if it is too high the gas flow at the cylinder should be

FIGURE 2.5: *Gas regulator. A gas regulator controls the CO_2 flow from the cylinder towards the incubator. There is a flow valve in the centre of the regulator; the left dial gives the pressure in the gas line to the incubator, while the right dial shows the amount of CO_2 pressure in the cylinder.*

switched off before the gas supply line is disconnected, and allowed to vent. Only when the supply pressure has been set should the gas switch on the incubator be activated, following which the incubator should be left for a few hours to equilibrate. Once this is done the CO_2 dial on the incubator can be checked and the supply altered accordingly.

It is good practice to test the gas composition in the incubator. Gas analyzer kits can be purchased from general laboratory suppliers, for example the 'Fyrite kit' (Jencons) draws the atmosphere over testing crystals. The cheapest and easiest method, however, is to place culture medium at pH 7.0–7.2 in the incubator in an open flask, overnight. A red/purple color indicates too little CO_2, a yellow color indicates too much, and maintenance of the original color indicates the correct CO_2 level. An experienced operator can assess the pH surprisingly accurately by observing the color of the medium, but the novice should check the pH using a meter (this needs to be done fairly rapidly before a significant change occurs while the medium is outside the incubator).

2.4.3 Day-to-day running

You should get into the habit of checking regularly the incubator for proper functioning.

Gas cylinders: these should be replaced as soon as the internal pressure begins to drop. It is wise to have a spare cylinder at all times. Devices are available which can automatically switch from one cylinder to another to enable an uninterrupted supply of gas should a cylinder be exhausted.

CO_2 level: does the dial read the correct level when the incubator is not taking in gas? Remember that, after opening, the incubator will take in gas and the level may overshoot the set level for a short while before equilibrating.

Temperature: check using an internal thermometer, most conveniently placed on the internal door. Remember that after heavy use the temperature of the incubator will take a while to rise.

Water level: the tray base must be completely covered with distilled water in order to maintain the humidity level; check the water jacket if applicable.

Contamination: it is advisable to check the walls of the chamber and the water tray; discoloration in patches is relatively easy to detect. If these are found, the affected areas should be first cleaned with detergent and then swabbed with 70% IMS. Contamination in the water tray can be prevented by adding an antimicrobial agent such as Roccal II, a germicidal sanitizing agent (Jencons).

Gross contamination: you may need to shut down and clean the incubator. The CO_2 supply at the incubator should be switched off and the water removed from the base. The interior can then be cleaned, first with detergent and water, and then swabbed with 70% IMS. Once this is finished the CO_2 supply can be switched on and the incubator allowed to equilibrate. This may be sufficient to cure the contamination problem, but more radical measures may be necessary. A method for fumigation is described in Chapter 7 (Contamination).

2.5 Centrifuges

For most cell culture only low-speed centrifuges are required. For safety reasons it is important that either the whole rotor or individual buckets are sealable: this will contain any spill or breakage and can also prevent aerosol spread. At the very least, the centrifuge chamber should be totally sealed when the lid is closed. Avoid old-style centrifuges which draw air through the chamber and expel it along with possible aerosols. Fine control of braking is a desirable feature, particularly when isolating cells using separation media. A gentle braking action helps prevent disruption of the separated bands of cells. In most cases cells should be centrifuged at 20°C; nevertheless refrigeration is useful to avoid exposing cells to uncontrolled higher temperatures due to heat from the motor.

2.5.1 Hints on centrifuge use

Centrifugation is a traumatic procedure for cells, and attention to a few basic rules can minimize its effects:

● The correct speed should always be used: increasing the speed can damage the cells. In general, 150–200 g for 5–10 min is sufficient.
● Cells should not be centrifuged for longer than recommended, as damage can be sustained when cells are compacted at the bottom of a tube.
● Once the spin has finished the supernatant should be decanted and the cells resuspended immediately.
● Centrifuges should be cleaned out regularly, as the bottom of the buckets or tube adapters will collect debris which can build up to cause imbalance in the centrifuge, or potential contamination when tubes are transferred to the hood. Naturally all spills and tube breakages should be cleared immediately.
● Normal safe practices for the use of the centrifuge are described in more detail in reference [1].

2.6 Microscopes

It is convenient and preferable to have both an inverted microscope and a standard microscope within the cell culture laboratory, both with phase contrast optics.

An inverted microscope is invaluable for visualizing cell cultures *in situ*. Looking at cultures in this way will give an immediate idea of their health and growth. A standard microscope with a movable slide holder is needed for counting cells using a counting chamber.

To get the most out of a microscope with phase contrast optics it should be set up correctly. The condenser plate should be turned to No.0 and the × 20 objective swung into place. One of the eyepieces is then replaced with a telescope eyepiece (which should be supplied with the microscope) and the telescope is used to focus on the dark ring in the back lens of the objective. When the × 20 phase plate is moved into place, a bright ring will become apparent. The bright ring should be coincident with the dark ring; if it is not, then the two screws on the condenser plate are adjusted until this is achieved. The telescope eyepiece can now be replaced with the normal eyepiece. Finally, a green filter, which should be available with the microscope, is placed in front of the light source.

A good sharp phase image should be achieved. In certain circumstances the illumination system itself may need adjustment. The light source must be centered and the iris diaphragm should be shut down to approximately 30% of its maximum size. If you change the objective remember to change the phase condenser plate and check that the phase rings are coincident as above.

Microscopes have sophisticated optics, which should be kept scrupulously clean. Any spillage should be wiped up immediately and special lens tissue should be used regularly to clean the lenses; the coincidence of the phase rings should be checked regularly, as described above.

Further information about the use of microscopes can be found in the appropriate microscope manuals and in reference [2].

References

1. Ford, T.C. and Graham, J.M. (1991) *An Introduction to Centrifugation*. BIOS Scientific Publishers, Oxford.
2. Rawlins, D.J. (1992) *Light Microscopy*. BIOS Scientific Publishers, Oxford.

3 Aseptic Techniques

Aseptic techniques are best learnt by watching an expert at work. A written text is an unsuitable means of describing all the manipulations you are likely to carry out in the culture of animal cells. However, it is worth demonstrating some of the most common operations that you will encounter in later chapters of this book, and some of the more fundamental errors: these are illustrated in a series of photographs in this chapter.

3.1 The culture hood work area

It is essential that the culture hood is scrupulously clean; this is best achieved by regular cleaning with 70% (v/v) industrial methylated spirit (*Figure 3.1a*).

A hallmark of successful tissue culture is the absence of clutter in the hood while work is in progress (see *Figure 3.1b*). Some essential items may reside in the hood, for example, a plastic-covered rack for Universal tubes; the waste beaker, containing hypochlorite; some medium, warmed to 37°C; an automatic pipet filler and any items of culture plasticware to be used immediately. Other items such as packs of pipets or culture dishes or flasks, should be removed from the hood once an item has been selected and then placed in a handy position outside the hood. A trolley with several shelves is useful for holding stocks of plasticware and appropriate containers for decontamination or disposal should also be to hand.

Figure 3.1c shows an overcrowded hood. Many of the items can be kept outside the hood until needed. A cluttered hood poses a number of hazards: there is insufficient working area; the proper air flow across the work area, which maintains sterility (see Chapter 2), may be compromised and a high risk of contamination is caused by contact between different items. There are three important rules which should be adhered to at all times:
• Keep the number of items in the hood to a minimum.

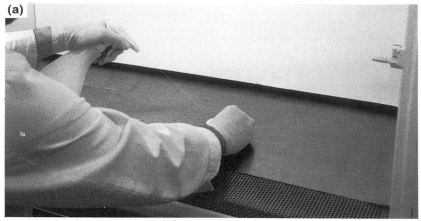

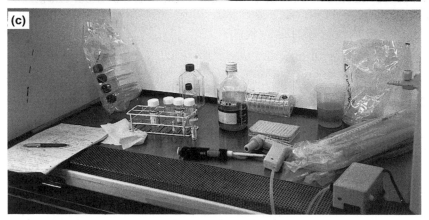

FIGURE 3.1: *The work area of a hood. (a) Cleaning, (b) tidy, (c) cluttered.*

- Remove all items once they are no longer required.
- Keep items well spaced, but easily accessible in the hood.

In a clutter-free hood, you should be working well inside the work area (*Figure 3.2a*). *Figure 3.2b* shows work being carried out in a non-sterile

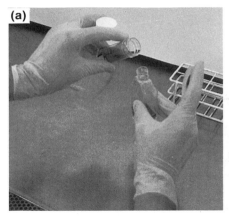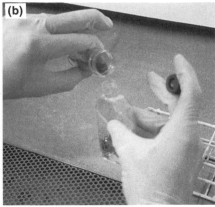

FIGURE 3.2: *The work area of a hood. (a) Sterile area, (b) non-sterile area.*

area of a Class II hood (see Chapter 2 for the air-flow diagram of such a hood). Similarly, in a laminar-flow (horizontal) hood, the air flow towards the front is affected by the air flow outside the hood, so work in this area may not be sterile. Working in these non-sterile areas is often the result of clutter towards the rear – this should be avoided at all times.

3.2 Handling pipets and automatic pipet tips

Sterile pipets should not be removed from a can in the manner shown in *Figure 3.3a*. Your hand will inevitably touch the sides of the can and many pipets other than the one selected, thereby possibly contaminating

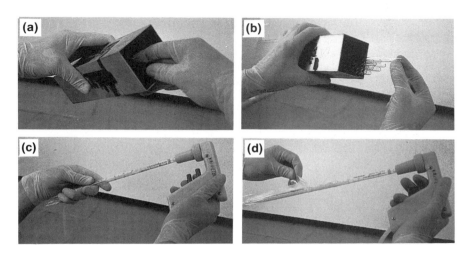

FIGURE 3.3: *Removing a sterile pipet from a can, incorrectly (a) and correctly (b); or from a plastic wrapper, incorrectly (c) and correctly (d).*

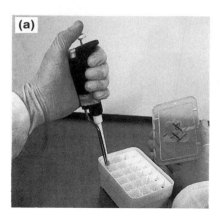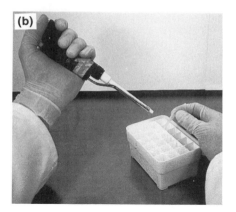

FIGURE 3.4: *Attaching a plastic tip to an automatic pipet; (a) correct, (b) incorrect.*

them. *Figure 3.3b* shows the correct way to access sterile pipets from a can, which is shaken gently, while angled slightly downwards, to move the pipets partially out of the can. A single pipet may then be selected without handling any other.

Plastic wrapping around commercial sterile pipets should not be removed in such a way that the body of the pipet is grasped by the hand (see *Figure 3.3c*). The pipet, and anything with which it comes into contact, may become contaminated. The correct way to remove the wrapping is to pull it back on itself, once the top has been torn and the pipet inserted into the automatic filler (*Figure 3.3d*).

To use a plastic micropipet tip from a rack, the lid should be lifted and the pipet pressed down into the tip before withdrawing it (see *Figure 3.4a*). Never remove the tip from the rack by hand and then press the tip on to the end of the pipet (*Figure 3.4b*) as this increases the risk of contamination.

3.3 Aliquoting medium from a sterile bottle, flask or tube

When the cap of the bottle has been loosened, it should be held between the fore- and middle fingers, so that when the hand is rotated to grasp the body of the bottle, the inside screw thread of the cap is facing away from the bottle (*Figure 3.5a* and *3.5b*). The lid should not be placed on the floor of the hood. This method reduces the risk of contamination by preventing contact of the inside of the cap with any potentially contaminated surface and by encouraging the operator to replace the cap whenever the bottle is

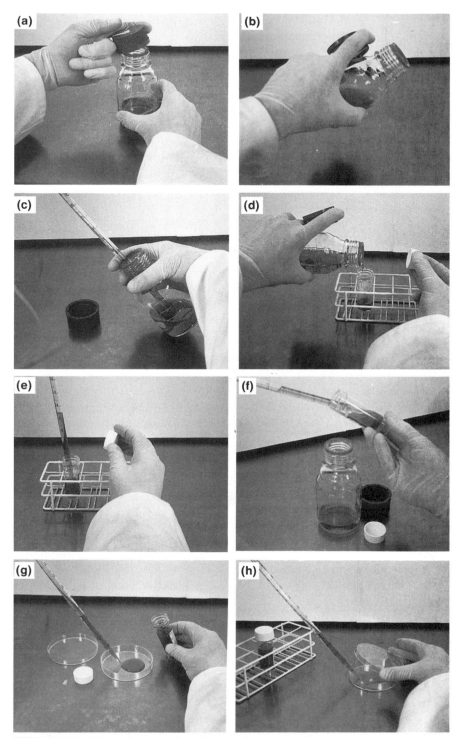

FIGURE 3.5: Aliquoting medium from a stock bottle to a culture dish; (**a, b**) removing and holding the cap; (**c**) incorrect aliquoting into pipet; (**d, e**) correct aliquoting into pipet; (**f**) incorrect handling of pipet; (**g**) incorrect transfer to dish; (**h**) correct transfer to dish.

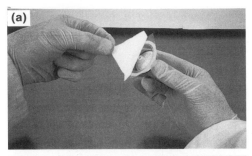

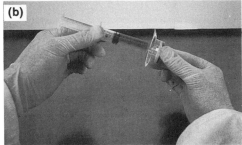

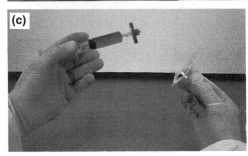

FIGURE 3.6: *Handling of syringe filter unit (**a–c**).*

not in use. At the same time it leaves the other hand free for other manipulations.

When removing an aliquot of medium, the bottle (tube or flask) should never be held by the neck (*Figure 3.5c*), only by the body (*Figure 3.5d*). *Figure 3.5c* also shows a pipet touching the side of the bottle (this should be avoided wherever possible) and a bottle cap incorrectly placed on the hood floor. Accessing media from a stock bottle using a pipet is best carried out in two stages. It should first be decanted into a smaller vessel (e.g. Universal) as shown in *Figure 3.5d* and then withdrawn into a pipet as in *Figure 3.5e*. This procedure reduces the risk of contaminating the entire bottle of a valuable medium.

Figure 3.5f shows two cardinal errors: failure to replace a cap on a bottle of medium and carrying out a subsequent manipulation above the bottle. Caps should be replaced on bottles immediately after use (but not necessarily screwed down) before a second manipulation is carried out. At a convenient point, when both hands are free, caps can be tightened.

Figure 3.5g and *3.5h*, respectively, show the incorrect and correct method for transferring medium to a culture dish. The medium should be taken up into the pipet (see *Figure 3.5e*), the cap on the Universal replaced and the culture dish lid lifted (not laid down).

3.4 Syringe filter units

Another common piece of plasticware is the commercially available syringe filter unit, for sterilizing small volumes. Once the protective covering has been peeled away, the syringe should be attached while the filter unit is in its plastic support (*Figure 3.6a–c*); the filter unit itself must on no account be handled.

4 Media Preparation

4.1 Requirements of cell culture media

In order to grow cells *in vitro* the environment needs to be as close as possible to that expected *in vivo*. Important environmental factors are the substratum upon which the cells grow, the medium in which cells are surrounded, and the temperature.

The complete formulation of a medium which sustains growth of mammalian cells in culture may seem very complex, and the number of different media available may appear daunting, but, in essence, any successful medium comprises an isotonic, buffered basal nutrient medium, providing inorganic salts, an energy source and amino acids, together with various supplements. For most cell lines the supplements are relatively few, the main one being serum or a serum replacement.

In this chapter guidelines are given to make up a general 'basic' medium to which further supplements are added to make it suitable for growing cells. Correct medium selection and preparation is an important part of cell culture. It is the difference in basic constituents, such as amino acids and salts and their concentration, which characterizes the different types of media.

4.1.1 Types of media and supplements

For most cell culture purposes a general medium type is perfectly adequate. Eagle's basal medium (BME) is one of the original defined media. This is commonly used for adherent (or monolayer) cells such as HeLa cells and some primary cells. Other media have subsequently been developed from BME, for example Dulbecco's modified Eagle's medium (DMEM). This was created by supplementing BME with increased amounts of vitamins and amino acids, and was first used for culturing embryonic mouse cells. However, it is now widely used for a broad spectrum of mammalian cell lines, particularly those grown as adherent cell monolayers. Minimum essential medium (MEM) was developed for

mammalian cells, grown as monolayers, with more fastidious requirements than BME could satisfy. It is widely used, particularly for primary
mammalian cells, but also for many established cell lines.

Additional Ca^{2+} and Mg^{2+} are included in many media preparations,
partly because these divalent cations are necessary for the adherence
proteins on monolayer cell types. Media for suspension cell culture do not
need these same high concentrations, for example **RPMI 1640** (Roswell
Park Memorial Institute 1640 medium) is commonly used for human/
murine normal and neoplastic white blood cells: these grow as suspension
cultures and have a reduced calcium requirement (see *Table 4.1* for
details of media components).

To support cell growth, basic medium requires the addition of supplements such as glutamine and serum. Glutamine is a major carbon source
for most cells in culture, providing precursors for further biosynthesis
and protein production. It also serves, in addition to glucose (and sometimes sodium pyruvate), as an energy source via the Krebs' cycle. Serum
contains an important but ill-defined mixture of growth-supporting compounds (e.g. polypeptides, hormones, lipids, trace metals). The components of serum which are important in aiding cell growth have been
analyzed, and this has permitted the production of a chemically defined
serum-like supplement. Its advantage is the elimination of interference
from unknown factors contributed by serum which could be detrimental
to some experiments. However, serum is still used by the majority of cell
culture workers, usually in the form of fetal calf serum.

4.1.2 Serum-free media

Serum-free media are becoming more widely used, particularly for
hybrid cells (see Chapter 11). **Iscoves** medium is often used as a base for
serum-free preparation; this is a modification of DMEM to which supplements (e.g. bovine serum albumin, transferrin and soybean lipid) are
added. Supplement kits can be obtained from companies such as Gibco.
Iscoves was developed for the growth of murine lymphocytes and hybrid
cells. Newer formulations of serum-free media include the **AIM V**
medium from Gibco and 'Serum-free and protein-free hybridoma media'
from Sigma. These can be purchased as complete formulations for certain
cell types and do not require further supplementation.

The use of serum-free media for a number of cell types requires the
addition of extra cell-specific supplements which, in many instances,
have not been fully determined. Although a cell type may grow in a
serum-free preparation, many other characteristics such as cloning efficiency may be deleteriously affected. Problems such as these may be
overcome by the use of serum replacement factor kits, which contain
many of the active components found in serum. These are now being

TABLE 4.1: Composition of three commonly used medium types

	RPMI 1640 (mg/l)	DMEM (mg/l)	MEM (mg/l)
$CaCl_2.2H_2O$	–	264	264
$Ca(NO_3)_2.4H_2O$	100	–	–
$Fe(NO_3)_3.9H_2O$	–	0.1	–
KCl	400	400	400
$MgSO_4.7H_2O$	100	200	200
NaCl	6000	6400	6800
$NaHCO_3$	2000	3700	2200
$NaH_2PO_4.2H_2O$	–	141	158
Na_2HPO_4	800	–	–
D-glucose	2000	4500	1000
Phenol red	5	15	10
Sodium pyruvate	–	110	–
L-arginine.HCl	–	84	126
L-asparagine.H_2O	50	–	–
L-aspartic acid	20	–	–
L-cystine	50	48	24
L-glutamic acid	20	–	–
L-glutamine	300	584	292
Glutathione	1	–	–
Glycine	10	30	–
L-histidine.HCl.H_2O	–	42	42
L-histidine	15	–	–
L-hydroxyproline	20	–	–
L-isoleucine	50	105	52
L-leucine	50	105	53
L-lysine.HCl	40	146	73
L-methionine	15	30	15
L-phenylalanine	15	66	33
L-proline	20	–	–
L-serine	30	42	–
L-threonine	20	95	48
L-tryptophan	5	16	10
L-tyrosine	20	72	36
L-valine	20	94	–
Biotin	0.2	–	–
D-Ca pantothenate	0.3	4	1
Choline chloride	3	4	1
Folic acid	1	4	1
i-inositol	35	7.2	2
Nicotinamide	1	4	1
Para-aminobenzoic acid	1	–	–
Pyridoxal.HCl	1	4	1
Riboflavine	0.2	0.4	0.1
Thiamine.HCl	1	4	1
Vitamin B_{12}	0.005	–	–

This table contrasts three different media, RPMI-1640 and DMEM being relatively rich and MEM being a more basic medium for less demanding cell types. Also note that these are basic formulations and many alternative versions of each are now available.

tailored to support either adherent or suspension cell growth, for example Ultroser G or Ultroser HY from Gibco. For the definitive selection of media you will need to refer to literature pertaining to the various cell lines, as there are many variants.

FIGURE 4.1: Examples of small-volume filter units available (courtesy of Sigma Chemical Co.).

4.1.3 Choices of media formulation: powder, liquid concentrate or ready made?

Most basal nutrient media are available in three forms – powder, 1 × liquid and 10 × liquid. A limited number of common media are also available as autoclavable powders. Different types of media can be prepared by the same basic methods, these are outlined in this chapter.

Powder medium is particularly convenient when making large amounts (e.g. 5–10 l) at one time, and is the cheapest option. This option is often used by a busy laboratory which does not have access to a central media-making facility, or for the production of a single batch of medium for a particular experiment. Powder medium, however, does require a positive-pressure filtration system. Smaller volumes of medium (0.5–1 l) can be rapidly made from powder and filtered using ready-sterile filter units with attached sterile bottles (*Figure 4.1*).

Liquid media in concentrated form (10 ×) is a flexible method of making up just one or many bottles of medium at one time without having to use a positive-pressure filter system. Careful attention to avoiding contamination is very important during dilution and aliquoting, and aseptic techniques must be used.

Ready to use (1 ×) medium is the simplest option but is expensive, although supplements (e.g. serum and glutamine) still need to be added before use.

4.2 Requirements for media production

- A minimal standard of water for media preparation is double-distilled, but preferably triple-distilled or de-ionized and then double-distilled. Modern filtration systems (e.g. Millipore, as shown in *Figure 4.2*) often include carbon filtration, high-quality de-ionization and filtration. This water is usually of a high enough quality for cell culture.
- The bottles used for medium should be of the type made by Schott or Pyrex (available from most tissue culture suppliers), with pouring rings and substantial screw caps over broad necks.
- The bottles should have no trace of detergent – glass must be washed using a free-rinsing detergent and involve extensive rinses in distilled/de-ionized water.
- A laminar-flow tissue culture hood with an open front is needed, particularly for powder media filtration, giving easy access to the working area.

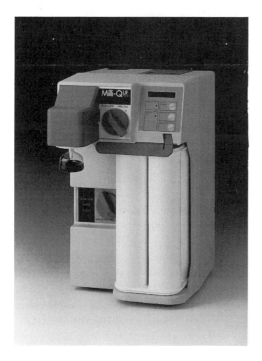

FIGURE 4.2: *A typical water purifier. This model is capable of producing water of the appropriate quality for media production (courtesy of Millipore Ltd.).*

TABLE 4.2: *Bicarbonate additions for various medium types*

Medium type	NaHCO₃	
	ml/l of 7.5% solution	g/l of powder
MEM with EBSS[a]	29.3	2.2
Medium 199	29.3	2.2
RPMI 1640	26.7	2.0
DMEM	49.3	3.7
Hams F10	16.0	1.2
MEM with HBSS[b]	4.7	0.35
Medium 199 with HBSS	4.7	0.35
BME with HBSS	4.7	0.35

[a] EBSS, Earle's balanced salt solution
[b] HBSS, Hank's balanced salt solution

4.3 Liquid media preparation

Bottles of pure water (see Section 4.2) are autoclaved, the caps being loosened beforehand and secured again once the process is complete (see Chapter 2). 1 M NaOH or 1 M HCl should be filtered with a 0.2 µm disposable sterile filter and stored in a sterile plastic container. An extra supply of sterile pure water should also be available. Bicarbonate solution for buffering is available commercially as a sterile 7.5% (w/v) solution, otherwise a solution at this concentration (using sodium hydrogen carbonate powder) can be made and filter-sterilized with a disposable 0.2 µm filter.

The appropriate amount of 10 × concentrated liquid should be added to the bottle containing water, e.g. 50 ml of concentrate to 400 ml of sterile water in a 500 ml capacity bottle. Swirl very gently to mix, and avoid contaminating the neck of the bottle.

Bicarbonate solution should be added to achieve a final concentration of 2.0 g/l (0.2%). This concentration may depend on the medium being prepared and should be checked beforehand with manufacturer's instructions, (see Table 4.2 for a guide). 1 M NaOH or 1 M HCl should now be added (in small amounts and the bottle swirled gently to mix) to take the pH to 7.0–7.2. An experienced operator can tell by eye the correct pH; alternatively an aliquot of medium (about 20 ml) should be transferred to a clear container and altered to pH 7.2 using a pH meter. The color of this medium is a standard against which the pH of the rest of the medium can be correctly adjusted.

At this stage pure water is added to make up to the complete volume, while taking into account the volume of glutamine and antibiotics still to be added. The medium can now be stored at 2–4°C, or the supplements may be added for ready-to-use medium, (see Section 4.6). The maximum storage times for media once made up are recommended by the manufacturers.

4.4 Media preparation from powder

4.4.1 Filter units

Powder media reconstitution requires sterilization by filtration, unless autoclavable powder medium is being used (see Section 4.5). Filter units and systems range from simple 'bottle-top' filters to positive-pressure systems comprising large-capacity (e.g. 10 l) pressure vessels and filters. The larger units are more expensive to set up initially but are convenient and cheaper in the long term. For smaller volumes, the bottle-top filters require only a water pump or electric pump (to generate negative pressure) for operation, and are therefore convenient but expensive.

Autoclavable filter units are now preferred: these have superseded the old tripod filter stands used for large volumes of medium. Make sure that the units are rinsed out after each use and resterilized immediately, to prevent any microbial growth.

4.4.2 Filtration method

A large container should be filled to 80–90% of the final desired volume with sterile double/triple-distilled or de-ionized water and stirred magnetically. While the water is stirring, slowly add the powder medium, rinsing out the packet with water. The medium should be stirred thoroughly until all powder has dispersed. Sodium bicarbonate should be added (as sodium hydrogen carbonate powder) – the amount can vary dependent on the medium being made up (see *Table 4.2*).

Once all the powder has fully dissolved the pH must be checked. Where a positive-pressure system of filtration is being used the pH should be adjusted to pH 7.0. This is 0.1–0.2 points lower than the pH finally required, because a positive-pressure system causes some CO_2 to be removed from the medium during the operation. Acid is required (as 1 M HCl) if the medium is very red. The solution should be stirred when adding the acid. The medium at this stage is non-sterile so a pH meter can be used directly to confirm the pH. Finally, pure water should be added, while continuing to stir, to make up the final volume required.

The sterile bottles should be positioned in the tissue culture hood, ready to receive the filtered medium. If nitrogen gas is being used as the pressure system (N_2 gas is inert) the medium can now be decanted into the pressure vessel and the filter unit should be attached to the pressure line from the vessel (*Figure 4.3a*). The gas line should be attached and the

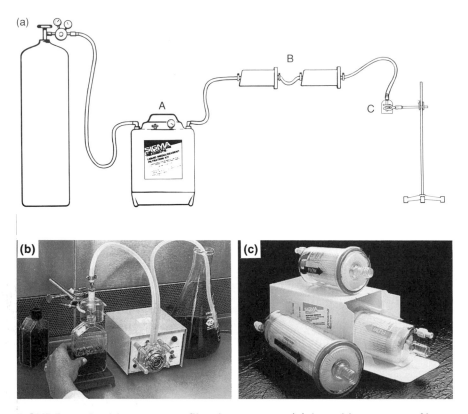

FIGURE 4.3: Positive pressure filtration systems. **(a)** A positive pressure N_2 gas system for filtering medium. The gas line leads to a pressure vessel (A) containing the medium which is forced (by the gas pressure) through the filter units (B) to a delivery device (C) which is positioned over a sterile bottle (courtesy of Sigma Chemical Co.). **(b)** A positive pressure filtration system using a pump to cause the flow of medium to the filter unit (courtesy of Millipore Ltd.). **(c)** Examples of capsule filters for use in volume filtration (courtesy of Sigma Chemical Co.).

gas maintained at a low pressure so that the flow of medium is at a controllable level. With a pump to provide the positive pressure, a set up can be used similar to that in *Figure 4.3b* which shows flasks being charged, although bottles can just as easily be used. Typical capsule filter units for use in medium filtration are shown in *Figure 4.3c*.

When filtration is complete, the pressure system is turned off and the pressure released from the pressure vessel. When dispensing, remember to allow sufficient volume for the subsequent addition of glutamine, serum and antibiotics.

A sterility check can be performed by putting the bottles of medium into a 37°C area overnight, and then they should be stored at 2–4°C. Before the medium is used, supplements need to be added as detailed in Section 4.6.

4.5 Autoclavable medium

Only a few types of media are available in an autoclavable form at present; this is made possible by removing heat-labile elements, which are then added after autoclaving. This is an alternative should positive-pressure filtering facilities not be available, but we recommend that all cell types are tested with the medium first: samples are available from most suppliers for testing.

The powder is dissolved in pure water and adjusted to pH 4.5 (necessary to stabilize the solution during autoclaving, but the exact pH should be checked with the instructions supplied with medium). The volume is made up (with the exception of the volume required for the buffer to be added after sterilization) with pure water, dispensed and autoclaved at 121°C, 103.5 kPa for 15 min. After sterilization the buffering solution/s are added aseptically and the pH of the medium is readjusted to pH 7.0–7.2. Refer to the manufacturer's instructions for the volumes.

4.6 Supplements

Having made the 'basic' medium, further supplements need to be added so that it can be used for cell culture.

L-*glutamine:* an amino acid which is an absolute requirement for cells in culture. It can be purchased as a ready-made sterile liquid (200 mM) or powder. The sterile solution should be aliquoted aseptically into 10 ml lots and frozen for future use. The powder should be made up with double/triple-distilled water to a 200 mM solution, filtered with a 0.2 µm disposable filter into 10 ml aliquots, and stored at −20°C. For use, the glutamine should be at a concentration of 2 mM, i.e. a 1:100 dilution of the 200 mM stock. Glutamine is labile in solution, with a relatively short half-life, especially at 37°C. After 3 or 4 days in an incubator 20% of the original glutamine may have broken down, even in the absence of cell growth. Medium which has had glutamine added should be stored routinely at 4°C, where more than 80% will still remain after 2 weeks.

Antibiotics: these are used by many laboratories but éschewed by some. Fastidious cells can be adversely affected by antibiotics, but gentamycin, penicillin and streptomycin are common additives in general cell culture. The usual active concentrations are: penicillin 100 units/ml; streptomycin 100 µg/ml; and gentamycin 50 µg/ml. Gentamycin is a broad-spectrum antibiotic and restricts the growth of mycoplasma as well as bacteria (Gram-positive and Gram-negative) and is therefore a convenient antibiotic to use (see *Table 4.3* for a guide to antibiotic use).

TABLE 4.3: *Details of some of the most commonly used antibiotics*

	Storage temp.	Life at 37°C	Active against	Working concentration	Mode of action
Gentamycin	4°C	5 days	Bacteria Gram +ve Gram −ve Mycoplasma	50 mg/l	Inhibits bacterial protein synthesis
Penicillin	−20°C	3 days	Bacteria Gram +ve	100 000 U/l	Interferes with bacterial cell wall synthesis
Streptomycin	−20°C	3 days	Bacteria Gram +ve Gram −ve	50–100 mg/l	Interferes with protein synthesis
Tylosin	−20°C	3 days	Bacteria Gram +ve Mycoplasma	6–10 mg/l	Interferes with protein synthesis
Nystatin	−20°C	2 days	Yeasts Moulds	10–200 U/ml	Permeability of cell membrane affected

Serum: serum is still widely used despite the rising popularity of serum-free medium. Typically, it is used at a concentration of 10% (v/v) but this does vary between cell types. Fetal calf serum is the most commonly used and is supplied to a high specification, though batch differences mean that it is good practice to test new serum for its ability to support cell growth. Other types of serum such as newborn calf, donor horse and human can also be used with some cell types.

Serum should be heat-inactivated before use to destroy the complement molecules and possible cross-reactive immunoglobulins which it contains. A complement cascade can result in lysis and the death of cells in culture. Inactivation is achieved by incubating the serum at 56°C for 30 min in a water bath. Serum is available from most suppliers either with or without heat inactivation.

Hepes: Hepes provides an alternative buffering system that can be used in a closed system, i.e. without CO_2. Hepes modified media can also be obtained commercially – these usually contain 25 mM Hepes; in some cases the salt (NaCl) concentration is reduced in order to allow the bicarbonate concentration to be maintained without compromising the osmotic balance of the medium. In non-gassed incubators bicarbonate will be lost from the culture as CO_2, this may adversely affect the growth of cells since buffering may not be the *only* function of bicarbonate in the medium. Hepes/bicarbonate combination buffered media can be used in a CO_2 gassed system when enhanced buffering capacity is required. Mops is another common organic buffer for use in non-gassed incubators at a 20 mM final concentration, in addition to the bicarbonate routinely added.

5 Culturing Continuous Cell Lines

5.1 Types of established cell line

Cell lines can be divided into two general types: adherent (monolayer cells) or non-adherent (suspension cells). Adherent cells attach to the plastic substratum of a flask or plate (*Figure 5.1a*) and therefore need to be detached from this surface before they can be used. Suspension cells, on the other hand, do not normally attach to the surface of the culture vessel (*Figure 5.1b*). The characteristics of cells in culture (*in vitro*) usually depend on their original source within the animal (*in vivo*). For instance, virtually all suspension cultures are derived from immune cells or precursors to immune cells [1,2]; *in vivo* they circulate within the bloodstream and do not tend to attach to available substrata (e.g. B- and T-lymphocytes). Adherent cells are derived from organs such as muscle, liver, nerve cells, kidney etc. and are not motile in the body.

Cells grown in culture, whether adherent or suspension, are often described as either primary, immortal or transformed. Primary cells (see Chapter 6) are those recently isolated from tissue or organs and usually have a limited lifespan in culture, sometimes called a finite cell culture. Both immortal and transformed cells have the capability for unlimited growth in culture (i.e. they do not die). Transformed cell lines are those cells which are either derived from tumor cells or have been manipulated in some way (for instance by transfection with oncogenes or treatment with carcinogens) to produce cells that display a novel transformed phenotype [3]. This transformed phenotype can manifest itself in a number of ways, such as newly acquired anchorage independence in adherent cells, reduced serum and growth factor requirements, or the ability to grow in soft agar and form colonies in immunosuppressed mice. Immortal lines are not necessarily malignantly transformed in the same way, but *are* altered to produce a continuously growing cell line from cells in a primary culture.

Although all cell types have similar basic requirements for growth, they can differ in the type of medium they require and it is vital that

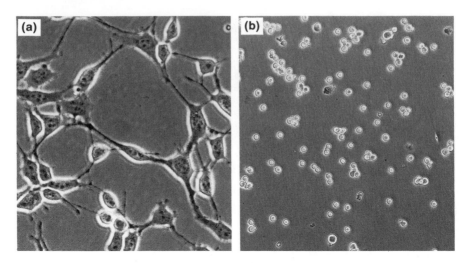

FIGURE 5.1: *(a) A culture of adherent cells (mouse 3T3 cells). (b) Cells grown in suspension culture (mouse L1210 lymphoma cells).*

appropriate nutrients are provided by using the correct medium (see Chapter 4). Commonly, RPMI 1640 is used for growth of suspension cells and DMEM for adherent lines; however, this is a very general guide. Certain laboratories use media on a purely 'historical' basis while others develop particular formulations of their own, using a common medium as a base. If you are unsure then refer to literature on the relevant cell type.

5.2 Preparation

A white coat, used for cell culture only, and gloves should be worn. The hood should be prepared and cleaned and the air pressure checked to confirm that it is performing to specification (see Chapter 2). Inspect the interior of the hood to check that it is clean. Swab the hood with 70% IMS – stubborn dirty areas can be cleaned with detergent and water followed by 70% IMS.

It is important to think ahead and assemble the required items prior to starting work: for example, culture medium and appropriate supplements should be warmed in an adjacent 37°C water bath, tissue culture plastics should be to hand and a clean hemocytometer ready to use. It is useful to have a small table or trolley, preferably with drawers, placed near the hood. This should contain all the items that may be needed – packs of sterile pipets, etc. – and helps keep the hood free of clutter.
• There is a huge range of tissue culture plasticware commercially available; a selection is shown in *Figure 5.2.*

Once the medium, serum and supplements have warmed then 'complete medium' can be made up ready for use as detailed in Chapter 4. For

FIGURE 5.2: *A range of plasticware available for cell culture. Standard cell culture flasks (1) can be scaled up to either roller bottles (2) or spinner culture flasks (3). Many different sizes of centrifuge tube are available, both conical and round-bottomed (4). Small-scale culture is usually performed in multiwell plates (5) of various sizes and can be scaled up to culture dishes (6). Media can be filtered with bottle-top filters (7) and small volumes filtered with sterile syringe filters (8). Also shown are plastic medium containers (9), and cryotubes (10) for freezing down cells. Not seen here are sterile Universal tubes, which are free-standing, polystyrene, screw-capped centrifuge tubes, that can hold up to 25 ml of liquid. Photograph courtesy of Bibby Sterilin.*

example, 500 ml of complete medium can be made by adding the following ingredients to 450 ml of basic medium: 5 ml of 200 mM glutamine, 50 µg/ml gentamycin and 50 ml fetal calf serum (to 10% (v/v) final concentration).

This is an example of a commonly used medium but it may vary according to the cell type or line used: for example, many cell lines can be grown adequately in 5% serum. Some laboratories use only penicillin and strep-tomycin, whereas others use no antibiotics at all. These differences can

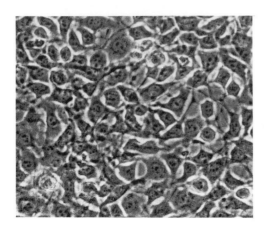

FIGURE 5.3: *A confluent culture of adherent cells.*

only be confirmed by referring to relevant literature about the cell line being used [4]. It is useful to perform cell growth studies using various concentrations of serum and other supplements if there is a question as to their use.

5.3 Assessing the cell cultures

On opening the incubator, you should observe the color of the medium. Yellow medium can indicate a CO_2 concentration which is too high, or overgrown cells (i.e. they have used up all the nutrients and exhausted the medium); if, in addition, the medium is cloudy, then contamination is probable (see Chapter 7 on Contamination). Conversely, deep red/purple medium indicates improper gassing of the cultures, either because the caps of the flasks were not loosened properly or because the CO_2 supply has run out (see Section 2.4.2 for further information on the CO_2 system.).

Adherent cultures should be examined under the inverted microscope with phase contrast optics (see Section 2.6) and a ×10 objective and phase ring to check the general health and confluence of the culture. Adherent cells should be spread as a monolayer on the plastic substratum. There may also be a few cells in suspension and a small number of cells stuck to the substratum but not spread out. The cells in suspension have just divided and have not yet found a place to settle down. Cells are said to be 'confluent' when no substratum can be seen between them and they are in physical contact with each other (*Figure 5.3*). Usually, cells are contact-inhibited and will stop growing at confluence. Subconfluent cultures are those where the cells are not in contact with their neighbors, with space for newly divided cells to adhere to. Normally cells should be subcultured (or 'split') before they become fully confluent. If cultures remain too long at confluence they may not grow well after subculturing; the reasons for this are unclear and still the subject of intense interest [5–7].

Suspension cells should normally not be seen adhering to the substratum (although in some cases they can be induced to do so) and healthy cells are rounded with smooth membranes, in contrast to adherent cells (*Figure 5.4*). Judgement of confluence in this case is more difficult and each cell type has differing requirements for optimum density of cell growth.

Construction of a growth curve (see Section 5.6) will enable familiarization with relative cell concentration and the corresponding confluency for both adherent and suspension cell types. Experiments often depend on how many cells are available and their concentration. In general, an experiment should not be initiated with fully confluent cells.

5.4 Counting cells

Before cells can be subcultured or used in an experiment, they must be counted. This is usually done with a hemocytometer with 'improved Neubauer' markings: this is a specialized microscope slide and coverslip. There are other automated ways of counting cells but these require specialized and expensive pieces of equipment such as Coulter counters; however, for most purposes the accuracy of the hemocytometer is more than adequate. Trypan blue is often used to estimate viability as it is excluded from the cytoplasm of viable cells; dead cells (even if they appear intact) are unable to exclude this dye and appear blue.

5.4.1 Preparing suspension cells

Ensure a homogeneous suspension by gently agitating the cells by tipping the flask back and forth and side to side. Sometimes the flask may need a gentle tap to loosen cells, which may have settled on to the substratum. A small volume (e.g. 100 µl) of the suspension should be removed with a sterile 1 ml pipet or sterile pasteur pipet and gently

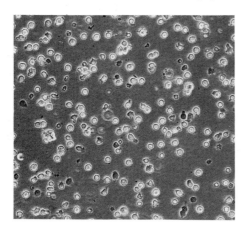

FIGURE 5.4: A confluent culture of suspension cells. Remember that different suspension cell types have markedly different relative levels of confluency and a growth curve should be constructed to assess this.

mixed with an equal volume of 0.5% (w/v) trypan blue in a small sample tube.

5.4.2 Preparing adherent cells

Adherent cells need to be detached from the surface of the tissue culture plastic before counting is possible. This is most easily achieved by using trypsin–EDTA (0.05% of 1:250 trypsin (i.e. trypsin which, under test conditions, can digest 250 g substrate for each 1 g trypsin added) and 0.02% (w/v) EDTA in balanced salt solution without calcium or magnesium). This reagent is also available as a ready-to-use sterile solution. It should be aseptically aliquoted into 10 or 20 ml volumes and frozen, but it should not be repeatedly freeze/thawed as this causes loss of activity. Adherence proteins need calcium and magnesium for their function, and therefore trypsin and EDTA are used in concert. The trypsin acts by digesting and thus cleaving the adherence proteins, and EDTA is present to chelate any free divalent cations. Trypsin activity is inhibited by the presence of serum proteins, which act as a target for the trypsin and which may also contain trypsin inhibitors. There are cases where the use of trypsin is not advisable, for instance when preparing cells for cell surface protein studies and/or antibody staining; should this be the case, alternative cell-removing strategies will have to be explored. This may be by mechanical means (e.g. cell scrapers) that are passed over the substratum of the culture vessel.

To release the cells from the substratum the medium is carefully poured off (or removed by pipet) and serum-free medium added (about 5 ml should be adequate) to the flask and washed over the growing area by gently rocking the flask. This medium should be removed by pipet and 2–3 ml (5 ml in a 75 cm^2 flask) of trypsin–EDTA added with a pipet and washed over the growing area as before. After removal of as much as possible of the trypsin–EDTA, the flask is placed in the incubator for 5 mins. Examination of the culture under the inverted microscope will show the cells to have rounded up, and many will have detached from the surface of the plastic. A gentle tap applied to the side of the flask should be sufficient to detach all the cells; 5–10 ml of complete medium should then be added from a pipet and the suspension mixed by gently pipeting up and down a few times. It is important to remember that trypsin is potentially damaging and should be in contact with the cells for the minimum time that allows removal of the cells from the plastic. Some of the cells may remain in clumps after trypsinization, especially if the culture was confluent. These clumps can be removed by transferring the cell suspension to a sterile 30 ml Universal container and allowing the clumps to settle; the supernatant can then be removed for counting. An aliquot of the suspension should be taken into a small tube to which is added an equal aliquot of 0.5% (w/v) trypan blue.

5.4.3 Using the hemocytometer

The hemocytometer is a thick glass slide with a central area isolated from the rest of the slide by two deep channels in the glass. This isolated area contains two counting chambers separated by a central reservoir (see *Figure 5.5a*). Etched on to the surface of the chambers are a series of lines as a grid (see *Figure 5.6*). When the coverslip is positioned across the central area, the space between the bottom of the coverslip and the top of the slide is a defined 0.1 mm. Each of the five squares, (the four corners and the centre), encloses 1 mm^2; this, combined with a 0.1 mm depth between the slide and the coverslip means that the volume above each square is 1 mm × 1 mm × 0.1 mm, i.e. 0.1 mm^3 (or 0.1 µl). Thus, once the number of cells in a square has been counted, the number of cells in 1 ml of the suspension is this value multiplied by 10^4.

Both counting chamber and coverslip should be clean and dry. The coverslip can then be positioned over the slide by placing 1–2 µl of water close to the edges of the channel away from the central chamber and pressing the coverslip down gently (*Figure 5.5b*). Make sure that enough pressure is applied to the sides of the coverslip so that it is firmly attached, but not so much that it breaks. 'Newton's rings', a rainbow effect between the coverslip and slide, can be seen if the coverslip has been secured properly. A sample of cell suspension (previously prepared as in Sections 5.4.1 or 5.4.2) is introduced into the counting chamber from a Pasteur pipet and placed at the edge of the coverslip over the central etched portion of the counting chamber. The suspension is drawn into the chamber by capillary action.

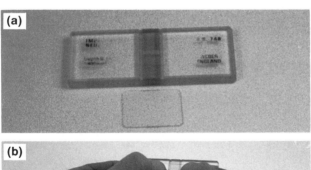

FIGURE 5.5: *The hemocytometer. (a) An 'improved Neubauer' counting chamber, complete with coverslip. (b) Fixing a coverslip to the counting chamber by pressing on the edges which have been moistened with a few microliters of water.*

(a)

(b)

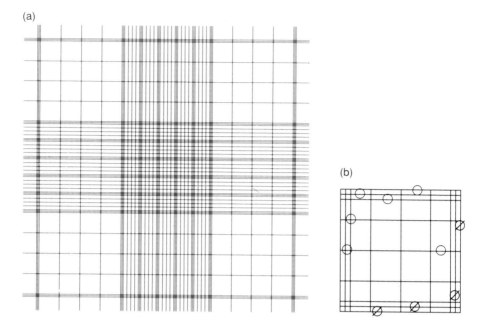

FIGURE 5.6: *Cell counting. (a) The grid of an 'improved Neubauer' counting chamber. There are five areas which can be counted. The central section has 25 small squares within an area of 0.1 mm². The four outer squares each have 16 squares within an area of 0.1 mm². All of these areas may be counted if necessary (courtesy of Sigma Chemical Co.). (b) An enlargement of a corner square indicating which cells should be included in the count (○) and those which should be ignored (⊘) (courtesy of Sigma Chemical Co.).*

The microscope objective lens should be lowered as far as possible, until it is just above the surface of the coverslip. Looking down the eyepiece the objective is then raised until the slide comes into focus. Using a × 20 objective and eyepieces you should be able to see the central square (*Figure 5.6a*). All the cells in the central large square should be counted, ignoring any cells that touch the outer tramlines on the bottom or right-hand side but including any cells that touch the same tramlines on the top and left (*Figure 5.6b*). Generally you should count 100, and preferably 200, cells in order to ensure an accurate count. If after counting the central square there are fewer than this, then those in any or all of the four outer squares should be counted until sufficient cells have been accumulated. The cell concentration per ml can be calculated as follows:

$$\text{cell count* } \times 2 \text{ (if trypan blue used)} \times 10^4$$

(*The count in one square or an average of a count if more than one square is used.) For example, a total of 210 cells were counted in three squares; this gives a cell density of $210/3 \times 10^4 = 7 \times 10^5$/ml.

In order to estimate the viability of the cell culture the count can be repeated without the addition of trypan blue and the two numbers compared to give a percentage of living cells. With experience it is possible to detect and count the non-viable cells without the use of trypan blue, since they appear darker under phase contrast than do healthy cells, but initially it is preferable to use trypan blue.

5.5 Subculturing cells

In general, suspension cells are diluted to concentrations of approximately 1×10^5/ml and should not be allowed to exceed approximately 8×10^5/ml, although this does vary between cell lines. For example, if a culture is calculated to be at a concentration of 7×10^5/ml viable cells, then it will need to be subcultured. To subculture into a 25 cm^2 flask using a 10 ml suspension of cells with an initial cell concentration of 1×10^5/ml, means that a total of 1×10^6 cells are needed. As the concentration of cells counted was 7×10^5/ml, then 1.43 ml of the culture is required and should be diluted up to 10 ml in fresh complete medium. For most cell lines this will be sufficient for a 3-day culture.

• It is informative and useful to compare the image seen under the inverted microscope and the cell count to grasp the idea of confluence and concentration.

Adherent cells are usually subcultured to a total number of 1×10^5 in a 90 mm dish, or 5×10^4 in a 25 cm^2 flask or equivalent; however, each cell line should be assessed individually. The growth area available is the most important factor here, but an adequate volume of medium must also be available for the cells to grow without its becoming exhausted. For example, after trypsinization and resuspension in 10 ml complete medium, a count has given a concentration of 1×10^5/ml, i.e. a total number of 1×10^6 cells.

To subculture cells into two 25 cm^2 flasks each containing 5×10^4 cells in a volume of 10 ml, 20 ml of cell suspension is required which contains 1×10^5 cells. However, it is always wise to make up a little more cell suspension than is necessary, therefore 1.1×10^5 cells should be taken from the counted cell suspension (i.e. 1.1 ml) and added to 20.9 ml of complete medium in a 30 ml Universal to achieve a total volume of 22 ml. After mixing, 10 ml of this cell suspension should be placed into each flask. *Table 5.1* gives some examples of the number of cells to seed into various culture vessels.

If using a CO_2 incubator the caps of tissue culture flasks should be loosened just enough to allow the gas to enter the flask. Conversely, when removing a flask from the incubator remember to secure the cap first.

If a dry non-gassed incubator is being used, tissue culture flasks should have their caps firmly screwed down to help prevent the evaporation of the medium. When the culture medium has warmed up to the incubator temperature it will expand and pressurize the flask, so excess pressure

TABLE 5.1: Guidelines for seeding number equivalents for various culture vessels, calculated relative to 1×10^5 cells/90mm dish

Vessel	Growth area cm^2	Adherent cell number $\times 10^5$	Medium volume	Suspension cell number $\times 10^5$
90 mm dish	49	1	10 ml	20–40
60 mm dish	21	0.4	5 ml	10–20
35 mm dish	8	0.15	3 ml	3–4
24-well plate	2	0.04	1 ml	1–2
12-well plate	4.5	0.09	2 ml	2–4
6-well plate	9.6	0.2	3–4 ml	3–6
25 cm^2 flask	25	0.5	5–10 ml	10–20
75 cm^2 flask	75	1.5	15–30 ml	30–60
162 cm^2 flask	162	3	30–50 ml	–

should be vented. An easy way to check for temperature equilibration is to observe when the initial condensation that occurs on the inside walls of the flask has disappeared.

5.6 Setting up a growth curve

It is important to construct a growth curve to assess the growth characteristics of the cells being cultured. They are simple to perform and provide information about the frequency of passaging, and also help in the prediction of cell numbers available for future needs. Setting up a growth curve involves the two techniques discussed above, counting and subculturing of cells. Suspension cells should be diluted to 1×10^5/ml, and subcultured into three separate flasks on day 0. Each of these cultures should be counted every day, for at least 5 days. The cell concentration and the time in hours should be recorded.

Adherent cells should be subcultured to 1×10^5 per 90 mm dish (60 mm dishes or 25 cm^2 flasks can also be used, but be sure to adjust the cell numbers according to the different growth areas). Three cultures are set up for each of 5 days, i.e. 15 separate cultures are initiated on day 0. This is because adherent cells need to be trypsinized and therefore for the purpose of counting each culture can be used only once. Again three cultures are counted per day for 5 days, and the cell concentration and time in hours are recorded. See *Figure 5.7* for guidelines on the setting-up of growth curves for adherent and suspension cells.

Growth curves can be constructed using standard graph paper but it is easier to use semilog graph paper (two- or three-cycle is sufficient). With semilog paper the cell number can be transferred to the graph directly. The cell number is plotted on the y-axis and the time elapsed since the initiation of the culture on the x-axis. The graph may show an initial lag phase where little growth occurs, then the culture concentration should increase in a logarithmic manner (exponential phase) which should be evident as a straight line. After a number of days the rate of increase of cell number slows and eventually increases no further (the plateau phase), and may even begin to decrease. *Figure 5.8* shows an example of a typical growth curve. The culture first enters the logarithmic phase at A and diverges from the log phase at B. At A the time in hours is C and the cell number is D. The cell number reaches $2 \times D$ at time E. The doubling time for the cells is therefore $E - C$.

The culture doubling time, once calculated for a cell line, allows prediction of the likely cell concentration at any time in the future. The point at which the growth curve leaves the exponential phase (B) is the point at which the cell growth rate decreases to a plateau phase. This can be due to exhaustion of the growth medium or the inhibition of growth through confluence. Cells should be subcultured *before* they reach a concentration at which their growth slows, which is why it is important to assess all cell lines by growth curves. The doubling time in cell culture is a characteristic for that cell under a particular set of growth conditions, and should not vary significantly.

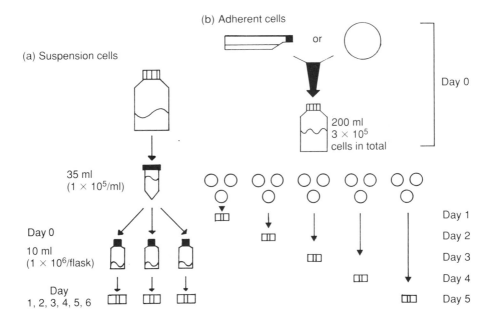

FIGURE 5.7: *Methodology for setting up a growth curve for suspension (a) and adherent (b) culture cells. Suspension cells can be repeatedly sampled from the three starter cultures, while for adherent cells three cultures are required for each day of counting. Thus, adherent cells must be plated out in multiple triplicates.*

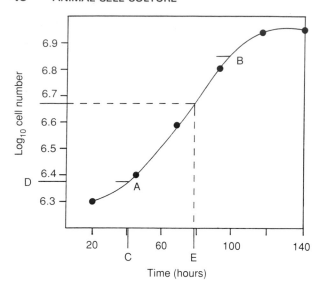

FIGURE 5.8: *A typical growth curve. See text for details.*

5.7 Scaling up production

Cell culture container sizes range from the very small (100 µl culture medium) to the very large (225 cm² growing area), the most popular being the 25 cm² flask, which usually holds 10 ml of culture medium, and the 75 cm² flask which can accommodate up to 50 ml. Larger tissue culture flasks containing over 50 ml of medium can have problems with efficiency of gas exchange and equilibration. Additionally, when suspension cells are cultured in a large volume, albeit at a low initial concentration, they will tend to settle on the bottom of the flask. This creates an artificially high local cell concentration in which some cell types will not thrive. Thus for suspension cells we advise against using any flask larger than the 75 cm²: these comfortably hold 50 ml of tissue culture medium, and gas exchange is not a problem.

There are a number of alternatives available with which to scale up cell culture; these include roller bottles and microcarrier beads for adherent cell cultures, and spinner flasks for suspension cultures. *Figure 5.2* shows some examples of larger culture vessels.

5.7.1 Roller bottles

In roller bottles the cells adhere to the total curved surface area, thereby markedly increasing the available growing area. These tissue culture bottles can be used in specialized CO_2 incubators with attachments that turn the bottles along the long axis. More usually, however, Hepes-buffered medium is used and the cultures are grown in a 37°C room. After each complete revolution of the bottle the entire cell monolayer has transiently been exposed to the medium. The volume of medium need only be sufficient to provide a shallow covering over the monolayer. The roller platform consists of a sturdy frame incorporating two or more moving belts upon which the bottles rest. The bottles should turn at approximately 2 r.p.h. In addition, it is most important to ensure that the platform is set horizontally; otherwise the cells at the higher end of the flask will die, as they are never bathed by the medium, which tends to pool toward the other end of the bottle.

5.7.2 Spinner cultures

Spinner cultures are used for scaling up the production of suspension cells. They consist of a straight-sided glass flask with a suspended central teflon paddle that turns and agitates the medium when placed on a magnetic stirrer. Commercial versions incorporate one or more side arms for sampling and/or decantation. An acceptable alternative is a Schott Duran bottle containing a sterile magnetic stirring flea. The cells are not allowed to settle to the bottom of the flask and thus cell crowding occurs only at very high densities, and stirring the medium improves gas exchange. However, as most incubators are not designed to accommodate the stirring plate, these are generally used in a warm (37°C) room, and the medium is buffered with Hepes (see Section 4.6). A speed of approximately 30 r.p.m. is optimal for these cultures.

5.7.3 Microcarrier beads

Beads are used to increase the number of adherent cells per flask volume, and are either dextran or glass-based and come in a range of densities and sizes. The beads are buoyant and should be used with spinner culture flasks. The surface area available for cell growth on these beads is huge (see *Figure 5.9*). As an example, Flow Laboratories's 'Superbead™ Micro-carriers', when resuspended at the recommended concentration provide 0.24 m^2 for every 100 ml of culture; this compares with $100-200 \text{ cm}^2$ achievable with the larger tissue culture flasks. Under these conditions adherent cells can be grown to very high densities before crowding becomes a problem. Cells growing at such high densities will rapidly exhaust the medium, which may need replacing during culture and therefore we recommend that Hepes is added to the medium to assist pH stability. New growth curves will need to be constructed for cells under these conditions, in much the same way as detailed before. The cells are removed from the beads by trypsinization in the same way as from a plastic flask substratum (see suppliers' recommendations).

These specialized cell culture accessories are available from a number of tissue culture suppliers such as Flow Laboratories, Sigma or Gibco (see

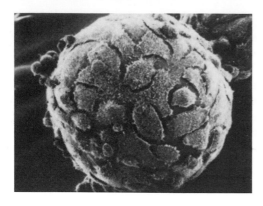

FIGURE 5.9: Growth of adherent cells on a microcarrier bead; note how many cells can be accommodated on one bead, magnification × 10 000 (courtesy of Sigma Chemical Co.).

Appendix B for further information). Detailed technical information on their use can be obtained from these suppliers.

References

1. Collins, S.J., Gallo, R.C. and Gallagher, R.E. (1977) *Nature*, **270**, 347.

2. Klein, E., Klein, G., Nadkarni, J.S., Nadkarni, J.J., Wigzell, H. and Clifford P. (1968) *Cancer Res.*, **28**, 1300.

3. Moore, G.E., Sandberg, A.A. and Ulrich, K. (1966) *J. Natl. Cancer Inst.*, **36**, 405.

4. Cell lines, and growth media used for cells supplied by Flow Laboratories, in *Flow Laboratories Product Catalogue*, p2.

5. Reiners, J.J., Pavone, A., Cantu, A.R., Auerbach, C. and Malkinson, A.M. (1992) *Biochem. Biophys. Res. Commun.*, **183**, 193.

6. Aoki, J., Umeda, M., Takio, K., Titani, K., Utsumi, H., Sasaki, M. and Inoue, K. (1991) *J. Cell Biol.*, **115**, 1751.

7. Wieser, R.J., Schutz, S., Tschank, G., Thomas, H., Dienes, H-P. and Oesch, F. (1990) *J. Cell Biol.*, **111**, 2681.

6 Primary Culture

6.1 Why use primary cells?

Primary cells are often used in preference to established cell lines and are regarded as a better representation of cells *in vivo*. It is felt by many that, in some cases, primary cells are more likely to reflect the true activity and functions that they display in their natural environment. What justification is there for regarding continuous cell lines as less useful than their primary cousins?

Many cell lines have been grown for periods in excess of 10 years: the biomass of some of these is now many times that of the original source organism. These cells often demonstrate significant differences from both the original culture and between cultures of the same cells in different laboratories. This divergence of phenotypes can cause problems in the interpretation and comparison of results. One example of this is the well documented observation that the rate of cell growth increases with prolonged periods of culture. This apparent adaptation to growth *in vitro* is often paralleled by the loss of characteristics associated with the original sample, and progressive changes in chromosome number. Compare, for example, the characteristics of the cell line HL-60 at the time of its original isolation [1,2] with the increased growth rate and reduced background of differentiation after 10 years in culture [3,4].

The major reason for the use of cell lines is that they are often easier to handle than primary cells, grow continuously, and large numbers of cells can be obtained. They are also readily available, and a relatively large amount of background information exists about them. Many cell types, however, are simply not available as cell lines: if there is no alternative, then considerations of ease of use and lifetime in culture are irrelevant.

6.2 Methods of isolation

Cells can be isolated from either whole blood, discrete organs (liver, heart, kidney etc.) or whole organisms (chicken embryos). It cannot be stressed too strongly that each tissue, organ or source organism presents its own particular problems, and methods of isolation must be tailored to each source. Such is the diversity of cell preparations and sources that to give protocols for all of them is impractical (an applications manual from Boehringer Mannheim, *Enzymes for Tissue Dissociation*, has a wealth of references and a number a useful protocols for cell types other than those described here; also see references [5] and [6]). Some sample methodologies are presented here which illustrate both the similarities and differences between methods. Although much of these methods is not strictly cell culture, they are procedures which require the background knowledge of cell culture gained from previous chapters.

• There will be local or national regulations governing the authorization for work with animals. These rules should be checked with your supervisor.

• Remember that, as with cell culture, every effort should be made to keep the sample sterile; all procedures should be carried out in a culture hood if available.

6.3 Isolation of chicken embryo fibroblasts

To achieve a suitable cell yield, 10–12 chicken embryos are required. Each fertile 10-day-incubated egg is first cleaned with a tissue soaked in 70% (v/v) ethanol. Then the broad end of the egg is opened using a pair of sterile scissors and the pieces of shell and white membrane underneath are discarded. The contents of the egg can either be poured into a sterile Petri dish or the embryos hooked out by placing No. 7 curved forceps between head and body. Then, using a pair of sterile forceps, the head is pinched off and the trunk transferred to a fresh Petri dish containing 5–10 ml of sterile calcium- and magnesium-free Hank's balanced salt solution (CMFH). The bodies are then washed twice in sterile CMFH at room temperature to remove the erythrocytes; this is most easily achieved by transferring to fresh dishes containing sterile CMFH.

With the embryo laid on its back, sterile No. 7 curved forceps are used to 'pinch out' the internal organs and breast tissue. Once this has been done and the internal organs discarded to waste, the remaining carcass should be washed three times with sterile CMFH as previously described. The embryo is now removed to a dry sterile tissue culture dish, and chopped into very small pieces with sterile scissors (nail scissors 10–12 cm with curved blades are best for this as they slide across the dish surface more easily). The tissue should be suspended in CMFH and kept on ice while other embryos are processed.

The pieces can now be transferred to a sterile conical glass flask (100–200 ml) containing 0.5 g of glass beads (710–1180 µm diameter, Sigma Chemical Co). Particular attention should be paid to the safety instructions on the container; glass beads are more dangerous than one would expect due to the powdered glass that can contaminate the beads. The glass beads may be suspended in 1 ml of PBS and autoclaved in the trypsinization flask before use. At this point 20–30 ml of ice cold 0.2% (v/v) trypsin (1:250) in CMFH can be added and the flask swirled at room temperature for 30 sec. After allowing the larger fragments to sediment for 30 sec, the supernatant containing contaminating erythrocytes can then be discarded. A further 15–20 ml of trypsin should now be added; swirled at room temperature for 30 sec; allowed to settle on ice for 30 sec and the liquid phase removed to a sterile Universal container. This should be repeated a further four to five times.
● Trypsin, glass beads and gentle agitation combine the elements of both enzymatic digestion and mechanical dispersion. Do not worry if small clumps of tissue are transferred at this stage.

The Universals should be centrifuged for 5 min at 4°C and 300 g, the supernatant discarded and the cells resuspended in 10 ml CMFH + 5% (v/v) fetal calf serum (FCS) or newborn calf serum (NCS) (this is to inhibit the action of the trypsin) and then centrifuged as before.
● Be careful not to use more severe centrifugation conditions to pellet the cells as this can result in reduced viable yield. Following resuspension in a total of 25 ml CMFH + 5% (v/v) FCS or NCS, the cells should be placed in a fresh sterile conical glass flask (with beads treated as before) and incubated on ice for 1 h, swirling every 10 min. This incubation allows any residual trypsin to continue the digestion, in conjunction with further mechanical dispersion.

After incubation the cell suspension should be poured through a cell strainer of 220 µm mesh (also called a screen cup, Sigma, see *Figure 6.1*) previously sterilized by autoclaving. The flask and beads are then rinsed twice in CMFH and the resulting suspension poured through the cell strainer. The cells should be pelleted as before and resuspended in 20–30 ml of DMEM + 10% (v/v) FCS (although many different basic culture media can be used for chicken fibroblasts). The cells can now be counted and plated out in either 90 mm cell culture dishes (1–3 × 10^5 per dish) or 75 cm^2 cell culture flasks (1–3 × 10^5 per flask). The cells should be examined under the inverted microscope the next day; the medium changed to remove any dead cells or erythrocytes, and if necessary subcultured by trypsinization as described in Section 5.5. Cell number can be altered according to the available culture vessels, but take care not to

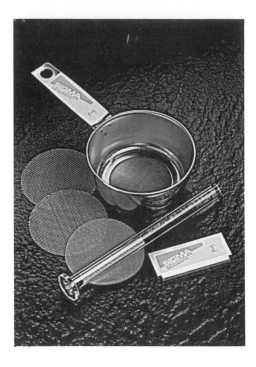

FIGURE 6.1: Screen cup filter. A screen cup which can filter tissue from the supernatant on the basis of the variable mesh size. The whole unit can be wrapped in foil and either autoclaved or sterilized in a hot-air oven (courtesy of Sigma Chemical Co).

provide too much medium as this can reduce endogenous growth factor concentrations, and impair efficient gas exchange in the incubator.

• Depending on the strain of chicken used these cells will grow for up to 30 passages; however, some specialized strains will grow for only a week before becoming senescent. *Figure 6.2* shows typical chicken fibroblasts.

• The digestion of tissue may be done with collagenase, or trypsin in combination with other agents such as Versene or ethylenediamine-

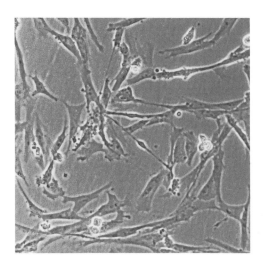

FIGURE 6.2: Chicken embryo fibroblasts 2 days after isolation from 10-day-old chicken embryos (cells courtesy of Esther Bell).

tetraacetic acid (EDTA), though the enzyme dispase is not recommended.
• Some workers prefer the cold trypsin method. The chopped pieces of
tissue are exposed to a 1:9 (v/v) mixture of 2.5% (v/v) trypsin (1:250) and
Versene (1:5000) (Gibco; Versene is an EDTA trade name available at
1: 5000 concentration) for 6–18 h at 4°C, this is then removed to a sterile
universal tube on ice. The tissue and residual trypsin/Versene is incubated at 37°C for 30 min, diluted with PBS, the lumps are allowed to settle
and the supernatant is transferred to sterile universal tubes.

6.4 Isolation of rat hepatocytes

Primary cultures of hepatocytes are often used in metabolic, pharmac-
ological and toxicological studies. However, the survival of such cultures
is relatively short (1 week or less) and it is necessary to prepare fresh cells
regularly. Attempts have recently been made to maintain primary hepatocyte cultures for 2–3 weeks by the addition of hormones to the medium
[7]. Alternatively, human hepatoma cell lines are available (e.g. Hep G2
cells); however, it must be remembered that these are transformed cell
lines and their metabolic and functional similarity to true hepatocytes
has often been questioned.

The liver consists principally of true hepatocytes (parenchymal cells) and
Kupffer cells (which form part of the reticuloendothelial system) together
with small numbers of endothelial and bile ductule cells. In order to
establish primary hepatocyte culture, it is first necessary to disaggregate
the liver by enzyme digestion followed by separation of hepatocytes from
other cell types. The method described below is for rat hepatocytes, which
is by far the most commonly used species. The method can be adapted for
other species and, in the case of large mammals, the caudate (small) lobe
of the liver is invariably used [8]. Most laboratory workers isolate hepatocytes by the two-step, non-recirculating perfusion method described by
Seglen [9]. In this method, the liver is perfused first with a Ca^{2+}-free
medium containing EDTA as Ca^{2+}-chelator. Since Ca^{2+} ions are involved
in cell–cell adhesions, removal of extracellular Ca^{2+} aids tissue disaggregation. The liver is then perfused with a second medium containing
collagenase and Ca^{2+} ions. Collagenase breaks down extracellular collagen, causing complete disaggregation of the liver and, paradoxically,
Ca^{2+} is required since collagenase is inactive in its absence. Many different preparations of collagenase are commercially available and it is
important to use a preparation specifically tested for hepatocyte isolation
(e.g. Sigma type IV collagenase).

Figure 6.3 illustrates a typical perfusion apparatus. The perfusion of rat
liver is a skilled operation for which the operator requires a license,
therefore this must not be attempted by untrained personnel. Towards

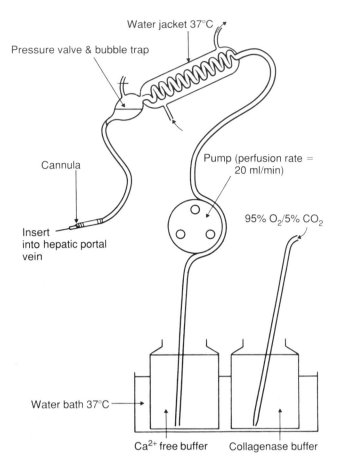

FIGURE 6.3: *Apparatus for rat liver perfusion.*

the end of the perfusion, evidence of cell detachment in the liver should be seen. The liver is now dissected from the abdominal cavity, placed in a small volume of collagenase-free Krebs–Ringer bicarbonate buffer containing 1 mM $CaCl_2$ and diced with scissors. The resultant suspension is gently stirred and filtered through 100 μm diameter gauze into a sterile universal container. Fibrous connective tissue and undigested pieces of liver are retained by the gauze.

The filtered cell suspension is then centrifuged at 50 g for 5 min. The supernatant (containing the majority of Kupffer cells, other cell types and non-viable hepatocytes) is removed by aspiration and the hepatocytes are washed twice by gently resuspending in collagenase-free Krebs–Ringer bicarbonate buffer, pH 7.4, containing 1 mM $CaCl_2$. Finally, the hepatocytes are washed once more into the chosen culture medium. Cell density and viability can be determined by mixing equal volumes of the hepatocyte suspension and 0.5% (w/v) trypan blue solution (in culture medium) and introducing an aliquot into the hemocytometer chamber (see Section 5.4.3). Non-viable hepatocytes appear blue with the nucleus staining particularly dark, while viable hepatocytes exclude the dye. A viability of >70% is usually required to effectively culture hepatocytes.

Many different media will support hepatocyte culture (e.g. Eagle's minimal essential medium, William's medium E and Waymouth's 721/1 medium). In order to initiate culture, the medium is usually supplemented with 10% (v/v) fetal calf serum together with antibiotics (e.g. 100 units/ml penicillin and 100 µg/ml streptomycin), amino acids (e.g. 2mM L-glutamine) and, in some cases, hormones. Hepatocytes are usually grown as monolayers on collagen-coated coverslips or in collagen-coated wells/flasks which are seeded with the appropriate volume of hepatocyte suspension at a density of 5×10^5 cells/ml. Cultures should be incubated at 37°C in a 5% CO_2 –humidified incubator. Viable hepatocytes attach within 3–10 h, after which the medium should be changed. If the hepatocytes are to be cultured for longer than 24 h, the medium should be changed daily.

6.5 Human primary cells

Some human cell types are readily available; for instance, human blood can be obtained either from volunteers (lymphocytes and monocytes), or umbilical cord (cord lymphocytes, venous endothelial and smooth muscle cells). Other cell types are sometimes available, such as human foreskin (fibroblasts), skin (fibroblasts) and tonsils (lymphocytes). However, many cell types will be impossible to obtain except under rare circumstances (e.g. normal liver, kidney, spleen, thyroid, pancreas). A compromise that allows the isolation of normal cells from organs is often to attempt to select these cells from the margins of tumorous or diseased tissue. This approach has been used to isolate cells from human thyroid tissue for culture and immortalization studies.

For more information of the use of human tissue for the isolation of primary cells see references [10] and [11].
• When handling human tissue the appropriate safety precautions must be taken. The work must be performed in a Class II cell culture hood.

6.5.1 Isolation of thyrocytes

• Human tissue is obtained from patients with multinodular goiter or Graves' disease undergoing partial or total thyroidectomy.

The sample should be placed in a sterile cell culture dish; any fat or connective tissue is cut off with sterile scissors and normal tissue removed from the sample. Normal thyroids are quite discrete and extraneous tissue easy to identify. Diseased tissue may present more of a problem and you should seek advice when separating this from the normal tissue at the margin. The normal tissue is chopped roughly with fresh sterile curved scissors, and then chopped finely on a McIlwain tissue chopper (see *Figure 6.4*) and the fine pieces (1–2 mm^3) placed in a pre-weighed sterile Universal container.

To every gram of tissue collected, 5 ml of 0.3% (w/v) dispase (Boehringer Mannheim grade II) in Ca- and Mg-free PBS (hereafter referred to as PBS) should be added and the container placed in a shaking water bath at 37°C. After 30 min the container is removed from the water bath and the tissue dispersed with a quill (a 5 inch filling tube, Universal Hospital Supplies Ltd, see *Figure 6.5*) fixed to a 20 ml syringe. The tissue can be drawn into the quill and syringe and then forced back into the container to provide high shearing forces. The container should then be returned to the water bath.

After a further 30 min the container should be removed from the water-bath and the tissue disrupted using the quill and syringe. The resulting suspension is then poured through a sterile (autoclaved) screen cup (220 μm mesh). The filtered suspension should be centrifuged at 500 g for 5 min and the pellet resuspended in 5H medium (see below). Because the

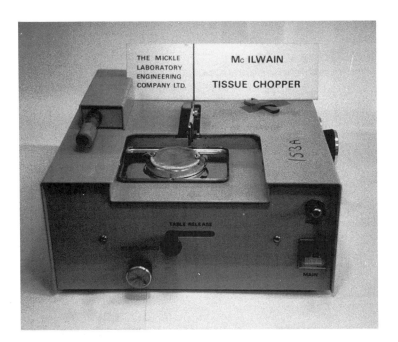

FIGURE 6.4: *A McIlwain tissue chopper. The whole unit can be placed in the hood, and the chopping plate swabbed with 70% (v/v) IMS. The razor blade (normal domestic design) is attached to the arm, which when engaged chops up and down whilst the table moves crosswise in increments. The table can be unlocked and turned as required to allow cutting across the previous line of chopping.*

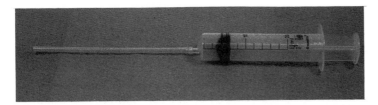

FIGURE 6.5: *A 'quill'. A 5 inch filling tube attached to a syringe, this 'quill' is used for mechanical disruption of tissue.*

cells tend to remain in clumps, counting them accurately can be extremely difficult. Once the cells can be counted, they should be plated out at 1.5×10^5 per 90 mm cell culture dish. These cells will be a mixture of thyrocytes and fibroblasts (if the surrounding tissue has been removed correctly the fibroblasts will be in the minority).

• The composition of 5H medium is as follows: Hams F10 + 5% (v/v) NCS supplemented with 10 µg/ml insulin, 10 ng/ml somatostatin, 10 ng/ml glycine-histidine-lysine acetate, 36 ng/ml hydrocortisone, and 5 µg/ml transferrin [12].

• As with chicken embryos the dispersion technique combines mechanical and enzymatic methods. The mechanical dispersion allows better access of the dispase enzyme to the tissue. Dispase is a non-specific protease effective against a wide range of connective tissue proteins which hold together adult tissue. An often used alternative is collagenase; however, many find that batch differences require each to be tested for effectiveness against the tissue of interest. Many workers also now recognize the value of multiple enzymes in crude enzyme preparations; DNAse contaminants in crude collagenase have been found to give better dispersion of enzyme-treated tissue than does the highly purified enzyme.

• Some workers recommend up to five digestions, each an hour long, in addition to mechanical disruption. The extent of the treatment is largely dependent upon the efficiency of the initial tissue chopping.

• Although these thyrocytes can be cultured for periods up to 6 months (albeit under very stringent conditions) they show very limited potential for growth. This illustrates one problem with primary culture, in that not all cell types appear to have the capability for growth *in vitro*. However, it is worth noting that it has been estimated that the size of the human thyroid can be accounted for by the proliferation of a single thyrocyte precursor cell and its daughter cells by only 5–10 divisions, and without abnormal stimulation these cells will not normally proliferate further.

6.5.2 Isolation of umbilical vein endothelial cells

The method described is based on that reported by Jaffe *et al.* [13].

• Human umbilical cords represent a relatively plentiful source of *fetal* cells; in this procedure, enzymatic treatment alone is used to strip off the cells that line the inside of the umbilical vein and are in contact with the fetal blood supply *in vivo*.

The cord should be placed in a sterile container with Ca- and Mg-free PBS and processing begun as soon as possible (if necessary it can be stored for up to 24 h in buffer at 4°C, though the viable cell yield will decrease accordingly). The cord should be inspected for any sign of trauma, i.e. clamping or damage to the outer surface. These areas should be discarded, as lesions on the interior surface will damage the basement layer under the endothelium and result in culture contamination with fetal smooth muscle cells. The cord has three vessels, two arterial and one venous: take care to use the correct one. Help may be required in the initial identification.

A three-way syringe tap (*Figure 6.6*) should be inserted in one cut end of the vein and the cord ligated to the tap with surgical thread. Using a 50 ml syringe the cord should be thoroughly perfused (washed out) with 100 ml of PBS. The other end of the cord is then clamped and 10 ml of 0.6% (w/v) dispase (Boehringer Mannheim grade II) in PBS introduced into the cord via the syringe tap. The cord is then transferred to a sterile plastic box containing enough PBS (equilibrated to 37°C) to partially cover it, and placed in an incubator for 15 min. The plastic box should be sterilized prior to use with 70% (v/v) IMS in water, and washed out with copious amounts of sterile water. After 7 min and at the end of the incubation the cord can be massaged gently, to increase the yield without causing smooth muscle contamination.

After incubation the cord is transferred to a sterile tray, the vein cut close to the clamp and washed out with 30 ml PBS via the syringe tap, and the outflow from the other end collected in a 50 ml Falcon tube. The suspension should be pelleted at 250 g for 5 min, the supernatant discarded and the cells resuspended in 20 ml of cord medium (Medium 199 mixed 1:1

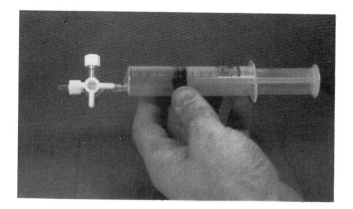

FIGURE 6.6: *Syringe tap. A three-way syringe tap used during manipulation of an umbilical cord. This allows the medium in the umbilical vein to be changed with the minimum trouble.*

with Earle's modified salts and supplemented with 10% (v/v) FCS or 10% NCS). After repeating the centrifugation, the cells are resuspended in 10 ml of cord medium supplemented with 20 μg/ml endothelial cell growth supplement (ECGS) and 15 units/ml heparin. The cells are then counted and plated out at a relatively high density: $5-7 \times 10^5$ cells per 25 cm^2 flask or 60 mm cell culture dish. To facilitate attachment of the cells, the plastic substratum must be precoated with gelatin (1% (w/v) gelatin in water is added to the dish, incubated at 37°C for 60 min, the excess removed and the dish allowed to air-dry; these can be stored at 4°C prior to use). After 24 h the cells should be examined under the inverted microscope and the medium changed to remove dead cells. The cells can be trypsinized as described in Section 5.4. They should grow with a doubling time of approximately 72 h and *can* proliferate for 3–4 months, although most workers use them within five passages, as they lose many differentiated functions extremely quickly. These cells can be cryo-preserved as described in Chapter 8.

• This method uses enzymatic digestion alone, in order to avoid contamination of the culture with smooth muscle cells. If smooth muscle cells are required then the cord vein can be clamped many times along its length, to disrupt the endothelium and allow access of the dispase to the smooth muscle cells.

• In this case the plastic substratum needs to be precoated to facilitate attachment of the cells. Additionally, note that growth factor and heparin are required and the cells are plated out at a relatively high cell density.

6.6 Special requirements for primary cell culture

Primary cell cultures are generally more fastidious in their growth requirements than established cell lines. This is probably due to the fact that many cell lines are derived from tumor tissue and may demonstrate reduced growth factor requirements.

The four examples of primary cell culture chosen above display quite different growth requirements. Some cells require little in the way of attachment factors or growth-promoting supplements (chicken embryo fibroblasts). Some cell types do not proliferate in culture (rat hepatocytes), although some survive for extended periods of time (human thyrocytes) and some cells proliferate for up to 4 months in culture, provided that they are supplied with attachment factors and specialized growth supplements (fetal endothelium).

The range of supplements required by different primary cells for satisfactory growth are legion and include the attachment factors, collagen (types I, II and IV), fibronectin and gelatin, which are derived from sources as diverse as rat tail and calf skin, and poly-L-lysine. Hormones and growth factors include insulin, insulin-like growth factor II (IGF-II), interleukin-2 (IL-2), interleukin-3 (IL-3), granulocyte/macrophage colony-stimulating factor (GM-CSF), endothelial cell growth supplement (ECGS), parathyroid hormone, fibroblast growth factor (FGF), epidermal growth factor (EGF), hydrocortisone, follicle-stimulating hormone (FSH), nerve growth factor (NGF), estrogen and testosterone. Each primary cell type requires its own particular cocktail of factors.

Human primary cell lines will grow for a variable but finite length of time in culture; after this time they become senescent and eventually die. In this respect human cells appear to be unique; cells from many other species, for example rodent primary cells, will also grow in culture and eventually become senescent. However, cell lines can often grow out of these cultures: they will grow indefinitely under the correct culture conditions and are thus immortal. The reason for this difference between species is unclear, although it has been speculated that the DNA repair and replication proofreading in the rodent is poorer than that of human cells. Thus rodent cell culture may reflect the multistep accumulation of genetic lesions that can result in a cell with a transformed phenotype. Chicken cells resemble human cells more than they do rodents, being more resistant to growing out of primary culture.

Sometimes primary cells survive but do not grow in culture. In the first instance this is unlikely to be the result of senescence, but may be due to the fact that the cell culture conditions do not suit them or, alternatively, the very nature and function of the cells *in vivo* does not include the ability to proliferate further (in other words they may be terminally differentiated).

6.7 Immortalization of primary cells

Recently, interest has focused upon the possibility of using transfection techniques to introduce so-called 'immortalization genes' into primary cells *in vitro*. Theoretically this could allow primary cells to proliferate as immortal cell lines, without loss of function or the expression of the transformed phenotype found in cell lines derived from malignant tissue.

In order to understand the concepts involved in the transfection of immortality genes, a little needs to be known about the lifecycle of their source organisms. One of the most popular of these genes is the SV40 large T gene (or T-antigen). SV40 is a small (5 kb genome) DNA virus of the Papovavirus family, which infects cells in an amphitropic way (i.e. across the species barrier). It was noticed that when SV40 infected quiescent cells (i.e. not proliferating) the synthesis of cellular DNA and cell replication was induced.

The induction of this growth in quiescent cells turned out to be crucial to the replication of SV40. These viruses lack the enzymes to replicate their own DNA and must therefore commandeer the host cell replication machinery. If the cell is quiescent this DNA synthetic apparatus is not active. Since *in vivo* the vast majority of cells are quiescent for most of the time, a virus is most likely to infect quiescent cells. However, the virus is equipped with the ability to force a cell out of quiescence (or G_0) and back into the cell cycle, where it will synthesize both host cell and viral DNA (in S phase). *Figure 6.7* illustrates the cell cycle of the host cell.

The ability of the virus to induce cell growth is central to the concept of the technique of transfection-induced immortalization. Many years of work in cell and molecular biology have now shown that individual viral genes, when cloned into plasmids and transfected into cells, can induce this cell cycle transition. Exactly how this is achieved is still the focus of intense research; one gene (SV40 large T) can, in some cases, induce proliferation in cells without changing their morphology (thus perhaps inducing immortality rather than transformation).

Plasmids expressing SV40 large T protein have now been used to immortalize a number of different primary cell types [11,12] in the hope that these cells, while remaining phenotypically normal, can be induced to grow continuously in a way that the normal cells cannot. For an explanation of some of the terms used here, see Chapter 14.

In many ways the initial promise of this new approach has failed to make the impact expected. One problem was found to be that cells which retain many of their *in vivo* characteristics in the quiescent state appear to lose

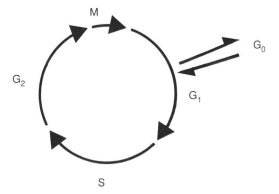

FIGURE 6.7: *The cell cycle. All cells reside at some point in this cycle. Quiescent (non-proliferating) cells are thought to be resident in G_0 (sometimes also termed G_0/G_1). With the correct stimulus the cells can move back into the cell cycle (probably at G_1) and progress through S phase (DNA synthesis) into G_2 (Gap 2) resulting in M (metaphase) followed by cell division and a return to either G_1 or G_0. Opinion is divided as to whether G_0 actually exists as a separate definable part of the cell cycle or is merely a convenient description of a cell held in G_1 for indeterminate time.*

these functions when they proliferate. Additionally, many of these immortal cell lines become genotypically abnormal: they either lose genetic material – sometimes a number of whole chromosomes – or they may almost double their chromosome number. Many of these problems may be unavoidable: the cells will become adapted to cell culture conditions and the better adapted cells, which grow fastest in culture, will be unerringly selected for by growth in culture. Also it has been suggested that the inappropriate growth of cells that are not normally required to traverse so many cell cycles may result in poor accuracy and proofreading of newly synthesized DNA. Therefore mutations may be introduced, resulting in cell transformation and a consequent change in the cell characteristics.

There is a growing suspicion that a cell *in vivo* can either grow *or* perform its other physiologically related functions, but not both at the same time. This theory is supported by the observation that some fetal cells grown in culture proliferate for longer than adult cells, but appear to be less differentiated. Many tumor cells isolated from adult tissue tend to be less differentiated than the associated non-malignant tissue; this is often termed de-differentiation of tumor cells, but may be no more than a consequence of their inappropriate growth.

In order to overcome this de-differentiation some workers have used temperature-sensitive mutants of SV40 large T [14]. This protein is fully functional at lower than normal temperatures (33°C) but cannot function when shifted to higher temperatures (37°C). Thus the cells proliferate at 33°C but withdraw from the cell cycle at 37°C. It is hoped that once the cells are quiescent they can more readily express the functions characteristic of their mortal, primary cousins.

Finally, many workers have come to the conclusion that cultured primary cells simply do not reflect the cells *in vivo* and any work can only be done effectively on cells freshly isolated from tissue. However, the process of tissue disruption alone can result in the loss of many specific cellular functions. All this reinforces the fact that data from any cell culture experiment, whether primary or established, must be interpreted with caution.

References

1. Collins, S.J., Gallo R.C. and Gallagher, R.E. (1977) *Nature*, **270**, 347.

2. Gallagher, R., Collins, S., Trujillo, J., McCredie, K., Ahearn, M., Tsai, S., Metzgar, G., Aulakh, G., Ting, F., Ruscetti, F. and Gallo, R. (1979) *Blood*, **54**, 713.

3. Macfarlane, D., Gailani, D. and Vann, K. (1988) *Br. J. Haematol.*, **68**, 291.

4. Schwartsmann, G., Pinedo, H.M. and Leyva, A. (1987) *Eur. J. Cancer Clin. Oncol.*, **23**, 739.

5. Freshney, R.I. (1990) in *Culture of Animal Cells.*, Wiley-Liss.

6. Jakoby, W.B. and Paston, I.H. (eds). (1988) *Cell Culture.*, Academic Press.

7. Dich, J., Vind, C. and Grunnet, N. (1988) *Hepatology.*, **8**, 39.

8. Emmisom, N., Agius, L. and Zammit, V.A. (1991) *Biochem. J.*, **274**, 21.

9. Seglen, P.O. (1976) *Methods Cell Biol.*, **13**, 29.

10. Whitley, G. StJ., Nussey, S.S. and Johnstone, A.P. (1987) *Mol. Cell. Endocrinol.*, **52**, 279.

11. Fickling, S.A., Tooze, J.A. and Whitley, G. StJ. (1992) *Exp. Cell Res.*, **201**, 517.

12. Ambesi-Impiombato, F.S., Parks, L.A.M. and Coon, H.G. (1980) *Proc. Natl. Acad. Sci. USA.*, **77**, 3455.

13. Jaffe, E.A., Nachman, R.L., Becker, C.G. and Minick, C.R. (1973) *J. Clin. Invest.*, **52**, 2745.

14. Jat, P.S. and Sharp, P.A. (1989) *Mol. Cell. Biol.* **9**, 1672.

7 Contamination

It is important to know what types of microbial organism may be present in order to devise the best course of action to deal with them. It is often a process of deduction to discover the source of the contamination, but it is best that this is done thoroughly in order to minimize future problems. In a multi-user facility it is difficult to isolate completely the cause of contamination, but good aseptic technique is a major way to reduce its incidence.

7.1 Types of contamination found in cell culture

It is essential that the agents of contamination – bacteria, fungi, yeast and mycoplasma – can be recognized (*Figure 7.1a*, *b* and *c*, respectively). Bacteria may be seen under the microscope as round bodies which exhibit Brownian motion, or they may be motile and rod-like. If they are not detected when the cultures are being scanned under the inverted micro-scope and prior to incubation, then the culture medium will become extremely acid (yellow) and cloudy. Bacteria will use up the nutrients in the medium and excrete waste products. Cultures with bacterial contamination must be discarded and care taken to prevent the spread of bacteria to other cultures, which should be checked for contamination. As a very general rule, bacterial contamination results from human touch, for example the operator accidentally touching a pipet tip, or touching the neck of a bottle of medium.

Fungi are particularly distinctive, with long hyphal growths giving, at the macroscopic level, a fluffy, fuzzy appearance. If the organism is detected before it becomes macroscopic it may be possible to prevent large-scale contamination: macroscopic growth means spores are being produced at a great rate. In general fungi spread in the air, and therefore it is difficult to prevent them from entering cultures once they have invaded a hood or incubator.

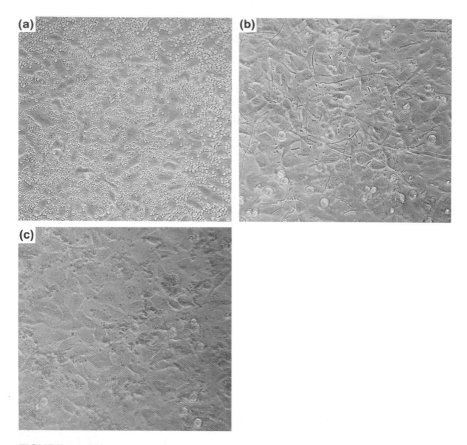

FIGURE 7.1: *Microscope visualization of contamination. (**a**) Yeast – note that the diameter of each yeast cell is 5–10 times smaller than the culture cells. Yeast may also appear oval or pear shaped and may also aggregate into short strings, rather like a pearl necklace. (**b**) Fungus – the first indication of fungus is likely to be long thin hyphal strands seen under the inverted microscope. Note the way in which these strands snake in and out of focus. (**c**) Bacteria – often difficult to see under the inverted microscope, and often mistaken for cell debris. In many cases the best indication of bacteria is strange clumps in the culture and a general increase in the debris associated with the cell culture.*

Spotting contamination at an early stage will often be too late for the contaminated culture, but may allow you to save other cultures that have not yet succumbed.

Yeast is a very uniform fungal microorganism (fungi imperfecta), with round or ovoid bodies that grow like a branched string of pearls. It tends to spread in a similar way to fungi. Home brewing or baking is often implicated in this type of contamination!

Mycoplasma (originally known as PPLO – pleuropneumonia-like organisms) differs from bacteria by being smaller, not having a cell wall and growing inside the contaminated cells. It is difficult to detect as it does not overtly affect the culture medium in which it is growing, and it cannot be seen by the light microscope. However, it may be possible to detect its effects on the cells being cultured, by observing whether they look 'normal' compared to the original cell type, and perform their expected functions in the usual way: for example, vacuoles in the cells are sometimes evident, or excess extracellular matrix can be seen. It is difficult to determine the origin of a mycoplasma infection; some investigators believe that it emanates from cell lines imported into the laboratory. Serum is a possible source but suppliers do state that their serum is free of mycoplasma; trypsin may be another source. It should also be remembered that mycoplasma can come from the mouth/throat of the operator as, like bacteria, it is widespread as a non-pathogenic organism. Once present in the laboratory, bad technique will cause it to spread further and seriously affect experiments by altering cellular functions.

7.2 Curing contamination

It may be fairly easy to isolate the cause of bacterial contamination if it has come from a single bottle of medium or serum. Disposal of the affected culture and other contaminating items can in many cases cure the problem. Fungal and yeast contamination present more difficulties because they spread via spore motility through air.

After a bad infection, fumigation of the incubator and hood may be required, but if the infection is isolated within a single plate or tissue culture flask it may be possible to avoid this. Before embarking on a full fumigation (see Section 7.4), and assuming the infection has been relatively light, it is wise to change the water and antimicrobial agent in the base of the incubator (see Section 2.4.2). Swabbing the surfaces of the incubator with 70% (v/v) IMS is also advisable. After the cultures have been removed, prior to cleaning, close down the caps and put them in a closed container during the short time they will be out of the incubator (do not risk contaminating another incubator). The medium should be checked by incubation at 37°C for 24–48 h and the hood thoroughly cleaned and checked for problems. If after this time fungi or yeast are still a problem, then fumigation must be considered (see Section 7.4).

Mycoplasma kits are now available with which cell lines can be tested; there are also laboratories which provide a test service. If there is a positive result the cultures should be discarded carefully. Consideration should also be given as to whether media or supplements may be affected: these should also be discarded if necessary.

7.3 Antimicrobial agents

Mycoplasma are inhibited by certain antibiotics such as gentamycin, although not all strains are susceptible. With the exception of genta-mycin, it is preferable not to grow cell lines with these antimicrobial agents on a permanent basis. Anti-PPLO agent (Gibco) and mycoplasma reduction agent or MRA (Flow Laboratories) are both claimed to cure con-taminated cultures, but they are probably best used as short-term pro-phylactic treatments with newly acquired cell lines, or those newly thawed from frozen.

Bacteria are best inhibited by broad-spectrum antibiotics, e.g. genta-mycin or penicillin/streptomycin. Whatever antibiotic is used, remember that some cells are susceptible to some antibiotic effects and that care should be taken when using more than one at the same time.

Inhibitory agents against fungi and yeasts are not recommended for long-term use as they can be toxic to cells – many will simply not grow in their presence. We would not recommend that these are used, except under exceptional circumstances.

7.4 Fumigation

Fumigation is not a pleasant job and should only be undertaken when absolutely necessary. Incubators need to be fumigated when the build-up of contaminants seriously hampers work. Tissue culture hoods, however, should rarely require fumigation except for servicing. The company responsible for maintenance often requires the fumigation of the hoods prior to servicing the equipment. The standard fumigation procedure is to use formaldehyde.

• *Important safety note:* formaldehyde is a toxic chemical that will cause burns in both the liquid and the vapor form. Therefore, wherever possible, dilution and dispensing of formaldehyde solution should be performed in a fume cupboard, and it should be transported to the hood or incubator in sealed vessels. Your local safety officer should be consulted regarding standard fumigation regulations in your institution and the testing procedures for formaldehyde. We recommend that breathing equipment is worn during these procedures.

Potassium permanganate (5 g) is placed in a 1 l glass beaker inside the hood or incubator (remember to switch off the gas supply at the incubator). To this is added 50 ml of formaldehyde solution (37–41% GPR). The combined solution will effervesce and give off noxious fumes;

the incubator or hood should be closed immediately and sealed with masking tape and plastic sheeting along any opening, to ensure that the seal is airtight. A warning message should be placed on the piece of equipment, so that it is not opened by accident.

After an overnight incubation, the beaker can be removed and its contents disposed of in a fume cupboard sink with copious amounts of water. A 50 ml beaker containing 5 g of ammonium carbonate is now placed in the equipment, which is resealed. This should be left for a further 2-4 h before removal. If the hood is vented to the atmosphere (i.e. outside the laboratory), then switching on the hood will vent the last traces of the formaldehyde. Incubators and some hoods cannot be vented in such a way, but residual formaldehyde should be totally eliminated before they are used again. More ammonium carbonate should be used and left overnight if any trace of formaldehyde remains.

8 Cryopreservation

The ability to cryopreserve cells is crucial. A cell culture worker will often need to use a variety of cell lines, but not all will be required to be grown at the same time. A large amount of time, effort and money would be wasted if cells could not be preserved when they are not needed. Cryopreservation (as the name suggests) involves the storing of cells at the very low temperature ($-180°C$) of liquid nitrogen in a state of suspended animation until they are needed. New cell lines and cell variants can be preserved as they were when originally isolated – this prevents possible problems of using cell lines that have become senescent or have lost their original functions. Samples of primary cells can be preserved against a future need not yet anticipated; contaminated cells can be discarded and a fresh batch of cells grown up.

- *Safety note:* Before attempting any work that involves contact with liquid nitrogen, it is imperative to ask an experienced member of staff or safety officer to demonstrate the safe placing and removal of cryotubes from liquid nitrogen tanks. Liquid nitrogen will cause burning due to its extreme low temperature. Treat liquid nitrogen with a great deal of caution, wear eye protection and heavy gloves.

8.1 Cryopreservatives

Healthy cells are suspended in a solution of either glycerol or dimethylsulfoxide (DMSO), the cryopreservative, with a high concentration of serum, cooled at a defined rate in liquid nitrogen vapor and then placed in liquid nitrogen. The function of the cryopreservative is to reduce the water content of the cells. The cryopreservative DMSO is a small molecule which is soluble in lipids, and thus enters cells extremely quickly by diffusion across the lipid bilayer of the plasma membrane. In the presence of DMSO, ice crystals, which would otherwise rupture cell membranes and cause them to lyze, do not form. The high serum concentration probably contributes to cell integrity by maintaining the intracellular protein concentration of cells rendered permeable by the DMSO. Glycerol has much the same effect as DMSO, and is used by some laboratories.

8.2 Freezing mixture

The most convenient method to add cryopreservative is to make up a double-concentrated freezing mixture as follows: 40% (v/v) growth medium (containing 10% serum), 40% (v/v) FCS and 20% (v/v) DMSO (or glycerol). The mixture should be made up in the order as stated and mixed well by inversion. The mixture can be filtered through a 0.2 μm sterile disposable filter into a sterile universal tube, and can be stored frozen at $-20°C$. When mixed with an equal volume of cell suspension in complete medium, the effective freezing mixture concentration is obtained.

8.3 Freezing down cells

Cells should be cryopreserved only when they are healthy and growing in exponential phase. Confluent or overgrown cells will not recover well from cryopreservation and therefore should not be used as a frozen stock. Cultures should also be checked carefully for contamination (see Chapter 7): there is no point in preserving contaminated stock. *Figure 8.1* highlights some of the following procedure.

The cells should be counted as previously described (see Section 5.4) and then centrifuged at 150–200 g at 4°C (if possible) for 5 min. The supernatant should be carefully decanted into a waste container, taking care not to disturb the pellet. The cells are then resuspended in the residual medium by gently tapping the side of the tube near the pellet, until no cells remain stuck to the bottom, and adjusted to *double* the final required concentration with fresh, ice-cold growth medium (antibiotic-free) containing 10% (v/v) FCS. As a guide, adherent cells should be resuspended to 2×10^6/ml and suspension cells to 1×10^7/ml. The cell suspension should be placed on ice and an equal volume of the freezing mixture added and mixed thoroughly. Aliquots (1 ml) of cell suspension should be put into cold prepared cryotubes (with a sterile 1 ml plastic pipet or micropipet). The final cell concentration is therefore 1×10^6 (adherent) and 5×10^6 (suspension) per vial. The screw caps of the cryotubes must not be overtightened as this will distort the gaskets and cause liquid nitrogen to enter the vials.

A freezing plug allows the cells to be cooled at a defined rate in the vapor that forms over the liquid nitrogen within the storage tank (*Figure 8.2*); the cooling rate should be 1°C/min. The cooling rate depends upon the type of plug and the number and position of the vials; instructions for

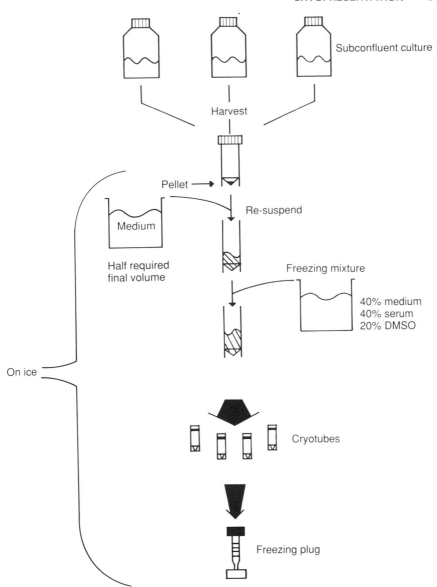

FIGURE 8.1: *A flow diagram illustrating the procedure for freezing cells. The cryotubes may be placed in a freezing plug as illustrated or alternatively placed in a polystyrene box at −70°C as described in the text.*

achieving the desired rate will be available from the manufacturers of the plug. Some plugs are lowered through the vapor at a controlled rate until the vials can be transferred to the liquid phase. If a freezing plug is not available, a polystyrene box is an acceptable alternative. It should hold 10–20 cryotubes with walls 5–10 mm thick and any dead space packed with tissue paper. The tubes should be placed in the box and supported to remain upright. The box should be tightly sealed and placed in a

−70°C freezer where the cells will cool at approximately 1°C/min and reach the temperature of liquid nitrogen vapor in approximately 3 h. At this point the cells can then be placed in the liquid nitrogen tank (*Figure 8.2*). Do not leave them in the −70°C freezer for longer than is necessary to achieve this temperature.

8.4 Thawing procedure

A water bath should be heated to 37°C and complete medium warmed up to temperature. The vial(s) should be collected from liquid nitrogen storage with care (liquid nitrogen can leak into the vial; when the vial is removed from the tank a rapid rise in temperature may cause the liquid nitrogen to squirt out of the vial). The vial must be thawed as quickly as possible in the 37°C water bath, following which it should be wiped with a tissue soaked in 70% (v/v) IMS before it is taken into the hood. The contents of the vial should be pipeted into a 10 ml sterile centrifuge tube containing 9 ml complete medium, the cap secured, and the contents mixed by inversion. The cells must be immediately removed from the freezing mixture in which they were stored by centrifugation at 150–200 g for 5 min and then resuspended in 10 ml of complete medium. A

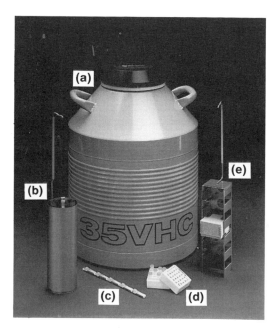

FIGURE 8.2: *A liquid nitrogen storage tank and accessories available (courtesy of Sigma Chemical Co.). (a) Liquid nitrogen storage vessel; (b) one type of holder which holds multiple 'canes'; (c) a 'cane' which can hold up to five cryotubes; (d) alternative system – a cryotube holder which fits into the holder labeled (e).*

sample should be taken for counting and to check for viability before incubation in an appropriate flask (or equivalent) at a subconfluent concentration for 24 h. The culture should be observed under the inverted microscope and counted again if necessary. Cells recover from liquid nitrogen with differing efficiencies and therefore no generalizations can be given, except to say that the cultures should be observed carefully for a number of days and subcultured when necessary. Some lines give a large percentage of dead cells after 24 h and take 3–4 days before the culture looks healthy and confluent, while others recover almost immediately.

Some cultures will be completely dead when recovered from storage, in which case another vial of the same batch should be tested and, if necessary, the whole batch discarded. The same advice applies to contaminated cultures.

9 Cloning Techniques

In cell culture terminology cloning refers to a method whereby a cell culture derived from a single parent cell can be obtained. Cloning is an extremely important part of cell culture and permits the selection of cell populations where each cell within the culture is identical to all the others (see *Figure 9.1*). Genetic changes will, however, occur during the growth of these clones and after a period of time the culture will no longer be clonal. In many cases this is not important, as cells are selected on the basis of a single characteristic which is under investigation: hence assays for this characteristic will monitor the continued suitability of the clone. There are many reasons to clone cells: for instance, a subset of a clone may diverge from the original phenotype, thus cloning allows the isolation of cells retaining the original characteristics. Other reasons for cloning cells are for cell fusion, immortalization and transfection which are covered in Chapters 11, 13 and 14. Here we will discuss what methods are available and in what circumstances they should be used.

9.1 Cloning by limiting dilution

Cells to be cloned are diluted and plated out into separate culture vessels in such a way that only one cell is contained in each vessel. Thus when a colony of cells appears it can be said to have originated from a single cell. Because the cell concentration is so massively decreased 'feeder cells' are often required to initiate cell growth and division. The type of feeder cell used is dependent on the cells being cloned. A further factor to be considered is the cloning efficiency of a particular cell type. This is a measure of the number of cells within a culture that have the ability to form colonies from a single isolated cell: this varies significantly between cell types.

9.1.1 Mathematical design

Limiting dilution cloning is most commonly performed in 96-well micro-well plates, each well containing 200 µl of medium. In practice the outer wells are not normally used as their contents tend to evaporate, therefore

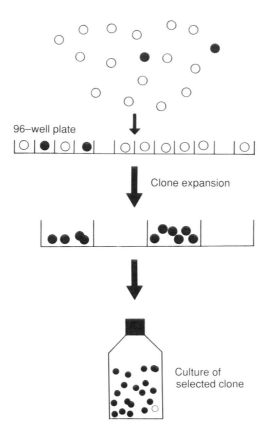

FIGURE 9.1: *Cloning. Cloning allows the selection of a single cell of interest (●) from a background of cells that are no longer required. Cells are plated in a 96-well plate at an average density of less than one per well. Choosing cells that grow in the presence of selective drugs (cloning and selection) or picking individual colonies of interest (screening of clones) and expanding those of interest will result in cultures of genetically (genotypically) identical cells (clones). However, these cells will not remain genotypically clonal (○) and will diverge as the culture expands.*

the inner 60 wells are used for the cell culture and sterile medium is added to the outer wells. The total cell suspension volume required for each plate arranged in this way is 12 ml. Cells to be cloned are diluted and plated at a concentration of 0.3 cells/well. Using these conditions only one-third of the wells would be expected to have one cell in them, and therefore the potential to form colonies. This increases the probability that any colony formed is derived from a single cell.

9.1.2 Feeder layers

One problem with the limiting dilution technique of cloning is that cells plated at low density often grow extremely poorly, if at all. This may be due to lack of conditioning of the medium and shortage of growth factors, lack of physical contact or the build-up of toxic cell debris. There are a number of ways in which this problem can be solved. Media suppliers now market a range of specialist tissue culture supplements that contain extra growth factors and nutrients to allow cells to grow at low density. However, the most commonly used technique has been the use of feeder layers. These are cultures of cells that do not have the ability to grow continuously in culture, and yet are able to provide the initial stimulus to the cells of interest by way of providing cell growth factors or cell contact, and they may also digest potentially toxic cell debris.

The type of feeder layer used is dependent on the type of cells being cloned. Macrophages are terminally differentiated, do not grow in culture, and will not overgrow the cells under selection. Other types of feeder layer are irradiated spleen cells, irradiated fibroblast cells and irradiated human peripheral blood cells for human cell lines. Irradiation causes cells to lose the ability to grow continuously in culture, but because they are not dead they can function in a feeder layer role.

9.1.3 Setting up limiting dilution cloning

Although the required dilution is 0.3 cells/well, it is recommended that other dilutions are also used, for example 1 cell/well and 10 cells/well. These represent huge dilutions of the original cell culture and so must be performed accurately.

Feeder layers, whether macrophages, spleen cells or human peripheral blood cells, should be prepared in complete medium (i.e. medium + 10% FCS) and irradiated with 30 Gy (3000 rads) of gamma-irradiation (usually achieved by exposing a cell culture to a gamma-emitting source such as ^{60}Co for a period of time which is specific to each source; consult your radiation safety officer prior to use). This is most easily done in a sterile tube (e.g. 30 ml Universal) and the cells should be subsequently plated out at 100 μl per well in the central 60 wells of 96-well plates. Macrophages should be plated at 2×10^4/well, while other cell types, such as spleen cells, should be used at $1–5 \times 10^5$/well.

• It is not within the scope of this book to describe the animal handling used in obtaining mouse peritoneal macrophages or other organs; you will therefore need to liaise with departmental or animal house members who are skilled in this technique (see also [1] for further guidance).

The cells to be cloned should be counted. It will be assumed in this example that the cells are at a concentration of 4.5×10^5/ml. The cells are diluted to a concentration of 1×10^6 cells in 10 ml of medium (1×10^5/ml) by adding 2.2 ml of the cell suspension to 7.8 ml of complete medium. *Figure 9.2* gives guidance on serial dilution. This suspension (100 μl) can then be added to 9.9 ml of medium; this is a 1:100 dilution and results in a cell concentration of 1×10^3/ml. A further 1:10 dilution (1 ml of cells plus 9 ml of complete medium) results in a cell concentration of 100 cells/ml.

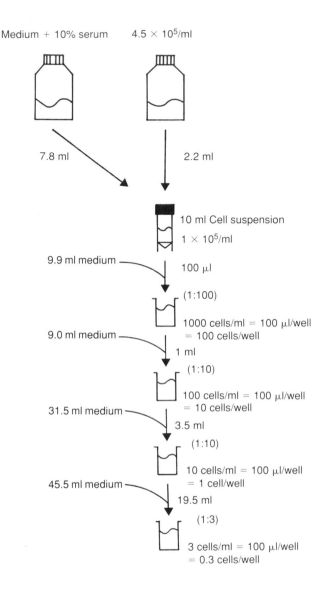

Medium + 10% serum 4.5×10^5/ml

7.8 ml

2.2 ml

10 ml Cell suspension
1×10^5/ml

9.9 ml medium

100 μl

(1:100)

1000 cells/ml = 100 μl/well
= 100 cells/well

9.0 ml medium

1 ml

(1:10)

100 cells/ml = 100 μl/well
= 10 cells/well

31.5 ml medium

3.5 ml

(1:10)

10 cells/ml = 100 μl/well
= 1 cell/well

45.5 ml medium

19.5 ml

(1:3)

3 cells/ml = 100 μl/well
= 0.3 cells/well

FIGURE 9.2: *Flow diagram of the procedure for setting up limiting dilution cloning. The scale of the dilutions is so huge that a serial dilution is the only way to ensure acceptable accuracy of plated cell number. Plating these different cell concentrations allows for the different cloning efficiencies seen with different cell types.*

When plated out in a 96-well microtiter plate at 100 μl/well, each well will contain 10 cells; one plate of this concentration will be required. Another 1 : 10 dilution gives a cell concentration of 10 cells/ml (3.5 ml plus 31.5 ml of medium). When plated out at 100 μl/well, each well will contain only 1 cell. The final dilution requires 19.5 ml of the 10 cells/ml dilution to be added to 45.5 ml of complete medium, resulting in a concentration of 3 cells/ml. Thus with 65 ml of this suspension a total of 10 plates can be prepared using 100 µl/well, which gives 0.3 cells/well.

These three cell concentrations should now be added at 100 µl/well to the 96-well plates previously prepared with the 100µl/l of feeder cells, giving a total volume per well of 200 µl. The addition of these volumes is most conveniently carried out using a multichannel micropipet. If a feeder layer has not been used, then the volume in the wells should be made up to 200 µl with medium. The plates are now ready to be placed in the incubator.

Subsequently 100 µl of the medium can be carefully removed once a week and replaced with fresh. Take care not to disturb the cells in the well too much, and avoid contaminating the contents of one well with another. Cross-contamination of wells will produce an inaccurate estimation of cloning efficiency.

9.1.4 Cloning efficiency

The number of plates prepared often depends on the plating or cloning efficiency of the cells to be cloned. In general one or two plates of 10 cells/well, two to four plates of 1 cell/well and four to 10 plates of 0.3 cells/well should be used. Cloning of this type is often repeated with a selected clone of interest to ensure that the required pure population is achieved. If the clone is unstable and begins to diverge from the original phenotype, then further cloning may be required.

Cloning efficiency varies between cell types, from as low as 0.1% to as high as 80–90%. It is important and useful to note the number of colonies appearing from each dilution: this gives an idea of the cloning efficiency of the cells, and also indicates whether or not a colony is likely to be derived from a single cell. For instance, if from the 1 cell/well plate only one well out of 100 produced a colony, then it is most likely that the colony is derived from a single cell. On the other hand if all wells produced colonies at this concentration, it is unlikely that any of the colonies are derived from a single cell.

9.1.5 Expanding the clone

As the colonies become apparent (depending on the cell type this will take 1–2 weeks), decisions concerning the next step need to be made. Since there are any number of reasons to clone cells it is not possible to comment upon the criteria used for decisions based on their subsequent processing. However, it will be assumed that at least one of the colonies needs to be processed further and expanded for cryopreservation. The secret of colony expansion is not to expand them too early or too quickly into a large volume of medium.

From a 96-well plate you may wish to expand each well to a 24-well plate

– each well can easily take 1 ml of medium. The other alternative is to go directly to a small tissue culture flask (25 cm²): when these are used in an upright position as little as 2 ml of medium can be used. This should not be attempted until the colony in the microwell plate covers 30–50% of the growing area, or fresh medium becomes exhausted within a few days of being added. One strategy is to remove half the cells in the microwell plate colony to expand, while maintaining the rest in the plate. Thus if one culture dies or becomes contaminated, the other should remain viable. Once an expanded clonal population is achieved, it is strongly recommended that they are cryopreserved as soon as there are sufficient numbers of cells.

9.2 Soft-agar cloning

9.2.1 Background

This is an extremely useful method which is sadly limited by the fact that some cell types do not grow in soft agar. The method avoids some of the problems associated with the limiting dilution type of cloning. It allows the worker to be confident that cell colonies are clonal while simultaneously enabling the cells to be plated out at a reasonably high density. This technique lends itself very well to a number of different selection and cloning protocols. It is usually used to select and clone from transformed cell lines in a single step. The authors have successfully used this technique to select and clone cells which have been transfected, are differentiation-resistant, and retrovirally infected, and those with metabolic deficiencies.

With soft-agar cloning (as with many other cell culture techniques) each cell line reacts in a unique way. Some cells will grow with a cloning efficiency of up to 90%, regardless of the concentration at which they were plated. Other cells are extremely dependent on concentration and only clone with a maximum efficiency of 10%, which can drop to as low as 0.1%. Preliminary cloning experiments should be performed before you attempt to select and clone cells by this method. *Table 9.1* gives some of the cloning characteristics that have been described:

9.2.2 Basic procedure

Here we will describe a procedure to determine the cloning efficiency of a cell type.
• The technique requires 'conditioned' medium, which is culture medium that has previously been used to grow cells for 2–3 days. In this case the medium should be conditioned by the cells that are to be cloned. The cells growing in this medium are assumed to have leaked

TABLE 9.1: *Cloning characteristics of cells in soft agar*

Cell line	Cell type	Cloning efficiency	Density dependent?
HL-60	Human promyelocyte	0.1–10%	Yes
K562	Human erythroid	50–60%	No
U937	Human monocytoid	NA	Yes
S49	Murine T-lymphocyte	80–100%	No

NA = Data not available.

growth factors which may then benefit the cells when suspended in soft agar during cloning. Conditioned medium can be collected over a period of time and stored frozen, or used fresh.

• In this sample procedure we will use RPMI 1640 medium and FCS.

• Experiments should always start with healthy subconfluent cells. They should be counted (see Chapter 5) and centrifuged: a total of 5×10^5 cells and 50 ml of conditioned medium are required.

The conditioned medium should have its serum concentration adjusted to 20% (v/v) (i.e. one-tenth volume of serum is added to RPMI + 10% (v/v) FCS). Adding to this an equal volume of fresh RPMI + 20% (v/v) FCS then gives 'half-conditioned' medium (HCM). 20% (v/v) FCS is generally used, as cells rarely grow in soft agar when 10% (v/v) serum is used.

The cells should be resuspended to a concentration of 2×10^5/ml in RPMI + 10% (v/v) FCS and subsequently further diluted with HCM to a concentration of 2×10^4/ml by adding 2.5 ml of cell suspension to 22.5 ml of HCM, i.e. a 1:10 dilution. This 1:10 dilution should be repeated four times to achieve cell concentrations of 2×10^3/ml, 2×10^2/ml, 20/ml and 2/ml. All the cell suspensions must be placed in a 37°C water bath for at least 30 min to equilibrate the temperature.

Twelve 60 mm diameter tissue culture dishes should be labeled with cell type, date and cell concentration. It is also a good idea to label the bottom of the dish in the same way as lids can get mixed up. 5% (w/v) Noble agar (which has been equilibrated to 60°C in a water bath) can be prepared by adding the Noble agar to distilled water and autoclaving. It is useful to do this in duplicate; the second one need not be sterile and can be used to monitor the temperature. The same approach can be used with a dummy cell suspension using medium alone. See *Figure 9.3*, which diagrammatically illustrates this procedure.

At this point a sufficient volume of 5% agar should be added to the cell suspensions to give a final concentration of 0.3% (taking into account the volume of the agar added). A simple formula permits the calculation of this volume for any dilution where the volume of the addition needs to be taken into account:

$$\frac{Va}{Va + Vm} = \frac{Cf}{Cs}$$

where Va = the volume of agar required; Vm = the volume of cell suspension to which the agar is added; Cf = the final percentage concentration of agar required and Cs = the starting percentage concentration of agar,

i.e.
$$\frac{Va}{Va + Vm} = \frac{0.3}{5}$$

Rearranging this:

$$1 + \left[\frac{Vm}{Va}\right] = \frac{5}{0.3} = 16.67$$

i.e.
$$Va = \frac{Vm}{16.67 - 1}$$

For all serial dilutions except the final one, $Vm = 22.5$ ml

i.e.
$$Va = \frac{22.5}{16.67 - 1} = \frac{22.5}{15.67} = 1.43$$

For the final dilution, $Vm = 25$ ml, i.e. 1.59 ml of agar is required for the 2 cell/ml suspension. It is important to note that the addition of 1.43 ml of 5% (v/v) agar to 22.5 ml of cells reduces the cell concentration from, for example, 2×10^3/ml to 1.88×10^3/ml. To avoid dealing with awkward numbers in the final suspension, the initial concentration of cells could be adjusted upwards to 2.127×10^3/ml before performing the same serial dilution.

Once the agar solution is removed from the 60°C bath it will start to gel as it cools: working with agar takes some practice and it may be a good idea to practice on some dummy cell suspensions first. A 5 ml pipet is used to transfer the agar to the cell suspension. The suspension is quickly, but gently, pipeted up and down two to three times to mix the agar with the medium before it can set. Capping the cell suspension and inverting it two to three times completes the mixing. If successful the agar will have gone completely into suspension. The 5% agar should be returned to the 60°C water bath, and the cell/0.3% (v/v) agar suspension to a 37°C water

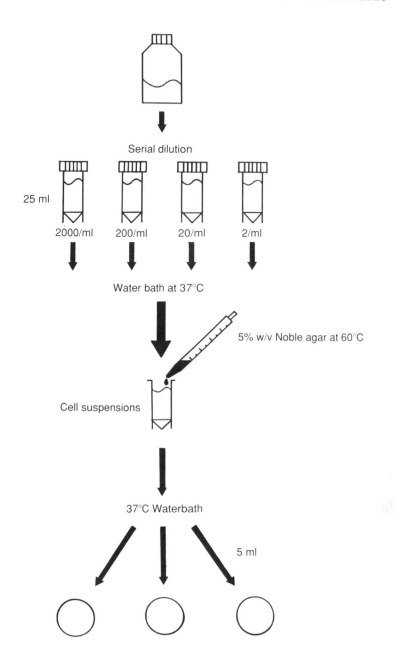

FIGURE 9.3: *Plating out serial dilutions for soft-agar cloning.*

bath. The process can now be repeated with the other cell suspensions, making sure the agar solution is re-equilibrated to 60°C between additions. A 5 ml pipet is then used to add 5 ml of cell/agar suspension to each 60 mm diameter plate (see *Figure 9.3*).

The agar plates should be left at room temperature for 30 min and then placed in the incubator. A number of the dishes should be observed under the inverted microscope to check that the cells are suspended singly. After 24 h the dishes should again be examined: some doublets of cells should be apparent. At this point a further 5 ml of RPMI + 20% (v/v) FCS should be added and after 3–4 days 2–3 ml of this medium should be replaced with 50% HCM. The old medium can be removed by carefully tipping the dish and slowly pipeting the medium from the top of the meniscus; the agar should remain attached to the bottom of the plate.

• Depending on the cell type, some evidence of colony formation should be visible to the naked eye after 1–3 weeks. *Figure 9.4* shows the typical appearance of soft-agar colonies. The time it takes for the colony to grow sufficiently to be visible will, of course, depend upon the growth rate of the cells. If and when they become apparent, the number of colonies can be counted and thus an estimate of the cloning efficiency can be made. It is probable that in the plates with the higher concentration of cells, counting the number of colonies will be difficult – there may even be so many of them that the

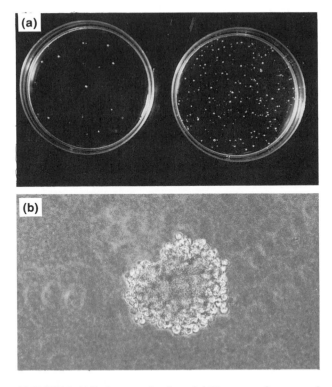

FIGURE 9.4: Soft-agar cloning. (a) Shows soft-agar colonies visible to the naked eye after 21 days in culture. Cells on the left were plated at 100 per plate and those on the right at 1000 per plate. (b) Individual colony in soft agar after 10 days in culture.

colonies will grow into each other. This is the reason for plating the cells at a range of concentrations.

9.2.3 Expanding the clone

• Soft-agar colonies can theoretically be picked as soon as they are visible, by either inverted microscopy or with the naked eye. However, it is our experience that colonies are best picked when they have attained a diameter of 1 mm. It is not easy to pick colonies from soft agar so we suggest practice with colonies used to determine the cloning efficiency of the cell line.

A glass Pasteur pipet that has been sterilized, either by autoclaving or baking in a hot-air oven 180°C for 60 min, should be used. The pipet is fixed to an automatic pipet filler and the end carefully pushed into the agar directly above the colony, continuing downward and enclosing the colony in the glass pipet. Applying gentle suction sucks the plug of agar containing the colony into the pipet. It is best to take as little of the medium as possible.

The colony can now be gently expelled (with the minimum of medium) into a culture vessel containing a small quantity of fresh medium + 20% (v/v) FCS. The culture vessel could be either a 96-well plate, 24-well plate, 12-well plate, or a 25 cm² tissue culture flask placed upright. A 24-well plate with each well containing 2 ml of medium, or a 25 cm² flask with 2-4 ml of medium, are the most convenient to use.

• The 24-well plate has the advantage that the cells can be observed with the inverted microscope; the flask has the advantage of the screw cap and thus may be easier to keep free from contamination.

Do not be tempted to disrupt the colony after it has been picked – it is better to allow the colony to seed itself into the surrounding medium. The colony should be allowed to grow without being expanded too quickly; the color of the medium and the crowding of the cells are useful features that can be used as a guide.

Cells should be counted when the culture volume is sufficient (i.e. using all the cells in the colony for counting would be pointless) and a minimum of five aliquots of approximately 5×10^6 cells should be frozen down as soon as possible. When possible, it is recommended that an additional batch of vials is frozen down.

• It is noticeable sometimes that the plates with the higher number of cells produce colonies both faster and with a proportionately higher cloning efficiency than cells at a lower cell number. Thus it appears that some cells have a density-dependent cloning efficiency.

• Soft-agar plates should not be disposed of too early, but if after 4 weeks no evidence of growth is seen then the conclusion must be that the cells are not cloning under these conditions. This may be because these cells simply do not grow in soft agar; alternatively, they may have an extremely low cloning efficiency. If this is the case they may clone if plated at a higher cell concentration, but if colony growth is not evident at 1×10^5/ml, then it may be time to accept defeat.

- In general, adherent cells are less likely to form colonies in soft agar than are suspension cells, and the more transformed the cell the higher the cloning efficiency. For example, in our experience EBV-immortalized B-cells (see Chapter 13) do not appear to clone in soft agar, and although they may sometimes appear to form colonies, this may be due to cell motility causing the cells to form clumps in the agar.

9.2.4 Alternative strategies

One problem with soft-agar cloning is that cells have a tendency to sink to the bottom of the plate while the agar is setting. This sometimes causes these cells to grow faster than the cells in the agar and overgrow the culture, exhausting the medium in the process. To avoid this a small volume of 0.5% (v/v) soft agar can be plated out first to cover the bottom of the well, and then the excess removed immediately. The cells in 0.3% (v/v) soft agar should then be plated out as normal. In order to stop the lower 0.5% (v/v) soft-agar layer drying out, the plate should be used immediately or stored in the incubator before use, but not for more than 30 min. This method allows the cells to be trapped in the lower agar layer.

Another modification is to prepare plates in advance with a confluent monolayer of irradiated feeder layer cells, and plate the soft-agar/cell suspension over these cells. A better alternative is to use conditioned medium from feeder layer cells for the medium changes.

Soft-agar cloning can be scaled up to 100 mm diameter plates using 20 ml of soft agar and 20 ml of medium, but the plate must be at least 20 mm deep (many are not); Falcon plates supplied by Becton-Dickinson are suitable. Cloning can also be scaled down to 12-well plate size using 1.5–2 ml agar/cell mixture and 2 ml of medium.

9.3 Ring cloning – cloning colonies from a single dish

For adherent cells, an alternative to limiting dilution cloning is ring cloning. This method has a number of advantages, notably less work and less tissue culture plasticware. It is often used to clone cells after transfection and drug selection. Cells before transfection are generally grown to a point 24 h before confluence: this maximizes the number of cells available for transfection but allows them to grow after treatment. Once transfected, the cells are incubated for 24 h then trypsinized and plated at a dilution of 1:5; the contents of one plate are now in five plates. Drug selection is now applied (see Chapter 14 for drug selection details) and over a period of time most of the cells will die; these lift off the plate and the debris can be removed during the routine medium changes. If the transfection has worked, the surviving transfected cells tend to grow as isolated colonies adhering to the substratum and, if the plates are disturbed as little as possible, the daughter cells tend to adhere closely to the rest of the colony. The plate will therefore contain a small number of individual colonies that remain to be subcultured as single colonies without contaminating one colony with elements of the other. Ring cloning is the method of choice in this case.

The equipment required for ring cloning is simple and can be home-made or it can be purchased from tissue culture suppliers. Rings can be fashioned from nylon or polyethylene tubing, preferably clear and about 5–10 mm in diameter. The tubing must be autoclavable – if you are unsure about the suitability of a particular type, sterilize a small piece to check. The tubing must be cut carefully into small lengths (about 10 mm) ensuring the cuts are as straight as possible. These rings can then be autoclaved in foil, autoclave bags or shallow wide-necked glass bottles. Some workers find that sterile silicone grease on the bottom of the rings will prevent leakage during subsequent manipulation. The commercially produced cloning rings have the advantage of a precision seal at the bottom of the ring and, although fairly expensive, can be reused many times.

The colonies to be cloned should be discrete and approximately 5 mm in diameter. The plate is prepared for trypsinization in the normal way, i.e. the medium is removed and the plate washed with sterile PBS (see Section 5.4.2). The colonies of interest should now be enclosed by the tubing rings with the aid of a pair of sterile forceps. Pipeting 100–200 µl of trypsin (as in Section 5.4.2) into the ring and then removing it immediately to waste provides sufficient trypsin, but does not result in the loss of cells. The plate can either be incubated at room temperature in the hood for 2–3 min, or, if the rings are thin enough, the lid can be replaced and moved to the incubator for 1–2 min. This procedure should be carried out without delay.

After incubation with the trypsin, 100–200 µl of medium with 10% (v/v) FCS should be added into the ring and the suspension pipeted up and down two to three times using a sterile Pasteur pipet. The cells should be placed in 1 ml of medium previously added to one well of a 24-well plate. This can be done for a number of clones simultaneously.

The secret of this technique is to start with isolated independent colonies, and not to attempt to remove *every* cell from the substratum. If a number of colonies are to be harvested from the same plate, then speed is more important than the efficient removal of the cells, because dehydration or over-trypsinization can kill the cells. Since there is no physical separation between the colonies when they are growing there is no guarantee that the cells in the colony all come from the same parent. For this reason it is wise to grow a substantial number of cells from each putative clone, and either perform a limiting dilution cloning or plate the cells at a low density and repeat the ring cloning.

Reference

1. Weir, D.M. (ed) (1986) *Handbook of Experimental Immunology*, Blackwell Scientific Publishers, Oxford.

10 Lymphocyte Separation and Guidelines for Establishing Lymphocyte Lines

This chapter covers the isolation of both human and mouse lymphocytes to provide cells for culture: for example primary cultures such as mouse T-cell lines, or a human immortalized B-cell line through EBV infection (see Chapter 13). Permanent cell lines can be created by the fusion of B or T tumor lines with their normal lymphocyte counterparts to create cells producing monoclonal antibodies or growth factors (see Chapter 11). The separated lymphocytes may also be used in assays such as cell killing, via natural killer (NK) or lymphokine-activated killer (LAK) cells (see Chapter 12).

The procedures recommended in this chapter should be seen as starting points from which to experiment and change as necessary, and not as definitive techniques. Once the basic techniques have been mastered, then alternative procedures can be attempted for optimal selection and growth of desired cell lines.

The choice of medium in which lymphocytes can be manipulated is wide. Some workers prefer to use serum-free medium to avoid the possible non-specific stimulation that may result from using fetal calf serum, whereas others use a common medium plus human serum for human lymphocytes. Different serum-free media are available from the major tissue culture suppliers; many are formulated especially for lymphocyte and hybridoma growth but they can be rather expensive. However, the use of DMEM or RPMI-1640 with FCS is still common in these techniques. The decision on which medium to use can only be made 'in house' and is often dependent on finances and convenience.

10.1 Isolation of lymphocytes

The first part of this procedure details the isolation of lymphocytes from human blood or tonsil tissue. Although we describe the teasing of lymphocytes from mouse immune organs such as spleen and lymph nodes, it is not the primary aim of this book to describe in detail the dissection of a mouse in order to obtain the immune organs. This is a skilled operation which should be performed by workers with appropriate experience, not a beginner. Consult your supervisor or animal work supervisor for advice on this matter.

10.1.1 From human blood

• *Safety note*: a Class II hood should be used and gloves worn at all times. Try to establish the infective status of the individual from whom the blood is derived. Care should be taken at all times to dispose correctly of all items that have been in contact with the blood and blood cells. Remember to have a waste tub containing hypochlorite in the hood before starting work.

The blood should be collected into a tube containing preservative-free heparin at a concentration of 10 IU/ml of blood and diluted with an equal volume of serum-free RPMI 1640 or equivalent medium.

Fractionation of the diluted blood is achieved by centrifuging through an equal volume of separation medium. This medium is a solution containing 9.6% (w/v) sodium metrizoate and 5.6% (w/v) Ficoll, which provides a density of 1.077 g/ml. Examples available commercially include Lymphoprep from Nycomed, Lymphocyte Separation Medium from Flow Laboratories, and Ficoll-hypaque from Pharmacia Biotechnology. The use of either 5 ml of separation medium in 10 ml sterile tubes or 10 ml in sterile Universal (25 ml) tubes is recommended. An equal volume of the blood/medium mixture should be slowly pipeted onto the separation medium (see *Figure 10.1a*). Alternatively, the separation medium can be laid under the blood using a syringe and metal cannula (see *Figure 10.1b*). The tubes are centrifuged for 20 min at 400 g (20°C). When transferring tubes between the hood and centrifuge it is important that the interface is not disturbed.

While the tubes are centrifuging, the caps of 10 ml sterile tubes are loosened, one for every separation tube being used. The lymphocytes drawn from the separation tubes will be put into these tubes.

Figure 10.2a shows how the lymphocytes band at the top of the separation medium. Using a Pasteur pipet the lymphocytes can be carefully removed from the interface, avoiding as much as possible of the upper layer and

(a) Overlay technique

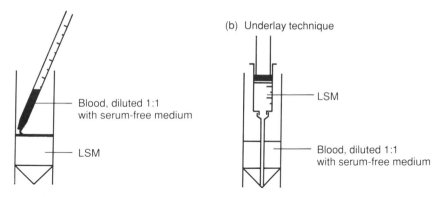

(b) Underlay technique

Blood, diluted 1:1
with serum-free medium

LSM

LSM

Blood, diluted 1:1
with serum-free medium

FIGURE 10.1: *Isolation of human lymphocytes. (**a**) The overlay technique. Blood diluted 1:1 with serum-free medium is slowly pipeted on to an equal volume of lymphocyte separation medium (LSM). The pipet should be held close to the top of the LSM and slowly raised as the diluted blood runs on to the LSM. This action should be executed as smoothly as possible. (**b**) The underlay technique. Blood diluted 1:1 with serum-free medium is added to a sterile tube. An equal volume of LSM is taken into a syringe with large-bore blunt needle, which is held vertically just above the bottom of the tube and the LSM slowly expelled with the minimum of turbulence created. Finally, the syringe and needle are carefully withdrawn.*

lymphocyte separation medium (*Figure 10.2b*). After placing the lymphocytes into the prepared tubes, these are filled with serum-free RPMI 1640. These tubes should then be centrifuged for 5 min at 200 g to pellet the lymphocytes, after which the supernatant can be carefully poured off and the pellet resuspended by tapping the base of the tube.

The lymphocytes should be washed twice by resuspension and centrifugation, using serum-free medium. A small sample of cells should be taken for a rough count before the last centrifugation. Finally, the lymphocytes are resuspended in complete medium to achieve a cell concentration greater than that required on the basis of the rough count previously made. A final count of the lymphocytes should be made and the volume adjusted to achieve the required concentration.

For separating large amounts of blood (i.e. greater than 30 ml) a preliminary step to remove the majority of erythrocytes is advisable. Use can be made of the ability of dextran to cause erythrocytes to clump and settle out of whole blood.

A stock dextran solution containing 3.5 g of dextran (mol. wt 250 000) and 0.9 g NaCl, is made up to 100 ml with pure water and sterilized by autoclaving. Dextran solution (24 ml) should be added to 30 ml of blood in a 50 ml tube. To hasten the separation this mixture can be incubated for 30 min at 37°C, after which time the majority of the red cells will have settled out. The supernatant should be pipeted off into fresh tubes and all

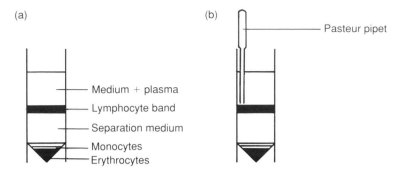

FIGURE 10.2: *Harvesting human lymphocytes. **(a)** Following centrifugation of diluted blood over a cushion of LSM for the recommended time a number of bands of cells should be visible. **(b)** The lymphocytes in the band at the top of the separation medium should be removed slowly and carefully by a Pasteur pipet. As little of the separation medium as possible should be removed when harvesting the lymphocytes.*

the cells counted. After centrifuging the cells at 200 g for 10 min they should be resuspended to a concentration of 4 × 10^5/ml in serum-free medium and then layered on to an equal volume of separation medium and treated as described above. More information on the separation of lymphocytes from human blood can be found in reference [1].

10.1.2 Lymphocytes from human tonsil

The organ is collected into a 50 ml tube (either with or without medium). When in the hood it is transferred to a 90 mm culture dish containing medium to rinse the tonsil and remove any connective tissue. After transfer to a fresh 90 mm dish containing about 5 ml medium, the tonsil is chopped finely with sterile scissors or scalpel. This suspension (including tissue lumps) is decanted into a new 50 ml tube, which is agitated and inverted to release lymphocytes from the tissue clumps. (Many lymphocytes will be obtained from this crude preparation, but if more are required then the tonsil can be pressed through nylon mesh as in Section 10.1.4.) When the clumps have settled the supernatant is transferred to a fresh tube for centrifugation at 150–200 g for 10 min. The cells are washed twice by resuspension and recentrifugation in serum-free medium before being resuspended in complete medium.

10.1.3 Mouse lymphocytes from spleen

The spleen should be placed in a sterile 60 mm culture dish containing 5 ml of serum-free medium. The spleen is gripped at one end with sterile fine forceps and, using sterile broad forceps, is squeezed progressively along its length, starting at the end held by the fine forceps. The empty 'skin' of the spleen should be removed from the dish to waste. This is a clean and effective way of obtaining spleen lymphocytes which can be mastered with practice. The cell suspension should be pipeted up and down using a 5 ml pipet to disaggregate any large clumps, and subsequently placed into a sterile tube and allowed to settle for 2–3 min. The resultant supernatant should be taken into a fresh tube and centrifuged for 7 min at 200 g (20°C). The lymphocytes should be centrifuged once more after resuspension in serum-free medium and counted.

● An optional step which uses ammonium chloride to lyze spleen erythrocytes can be added after the first centrifugation step of spleen cells. When the supernatant has been decanted, the pellet is resuspended in the small amount of remaining medium and then 5 ml of ammonium chloride buffer (144 mM ammonium chloride, 17 mM Tris pH 7.2, sterilized using a 0.2 μm filter) is added. After 5 min at room temperature, the supernatant should turn a clear red following lysis of the erythrocytes. The volume should be made up to 10 ml with serum-free medium before centrifuging the mixture at 200 g for 5 min.

10.1.4 Mouse lymphocytes from thymus and lymph nodes

The dissected thymus is removed to a sterile 60 mm culture dish containing 5 ml of serum-free medium and placed on a sterile nylon mesh (double or triple thickness of a 'wedding veil' type of net is ideal). Using a 5 ml syringe plunger gently squeeze the tissue – this will cause the cells to be released into the medium. The cell suspension should be pipeted into a separate dish and a 5 ml pipet used to aspirate the suspension up and down to disrupt any clumps. After transferring the cells to a sterile tube they are centrifuged for 5 min at 200 g. The lymphocyte pellet should be washed by resuspension and centrifugation twice more using serum-free medium. The lymphocytes can then be counted and resuspended in complete medium as required.

10.2 Enriching for T- and B-lymphocytes

There are numerous ways to select for T- and B-lymphocytes if the experiment requires a relatively pure T- or B-cell population. However, this is not always necessary and the need should be balanced by consideration of the effect of extra manipulation on the lymphocytes (contamination, poorer cell yield and decreased viability). It is recommended that a negative selection method is used to enrich, and although this will not produce an absolutely pure population it avoids possible stimulation of the desired cell type, which results from positive selection. *Figure 10.3* illustrates diagrammatically the strategy behind positive and negative selection. There are cheap traditional methods of enrichment which are described in Sections 10.2.1 and 10.2.2, but more sophisticated (and

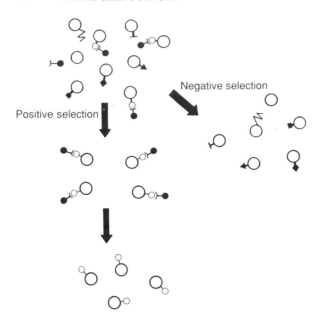

FIGURE 10.3: *The principle of positive and negative selection. Negative selection can enrich a population by removing a particular cell type and leaving others, whereas positive selection is more effective at isolating a particular cell type.*

expensive) methods, usually available in kit form, are summarized in Section 10.2.3.

Human lymphocytes consist approximately of 70% T- and 15% B-lymphocytes in blood, and therefore following a basic separation (Section 10.1.1) the human T-lymphocyte population is already partially enriched. If B-lymphocytes are required, it is possible to enrich the population by removing many of the T-lymphocytes.

Mouse lymphocytes can be selected by choosing a particular organ, for example, the spleen gives slightly more B- (40%) than T (35%)-lymphocytes, whereas the thymus and lymph nodes consist mainly of T-lymphocytes (>90%).

10.2.1 Enriching for human B-lymphocytes

AET rosetting is a quick, easy and cheap option which can give good results. The concept behind the technique is the utilization of the sheep erythrocyte receptor (CD2) found on human T-lymphocytes. 2-amino-ethylisothiouronium bromide (AET) in solution is used to treat sheep erythrocytes and the lymphocyte suspension is mixed with the treated sheep erythrocytes. The T-lymphocytes form rosettes around the erythrocytes and, following centrifugation through lymphocyte separation medium, the B-lymphocytes are left as an enriched population on top of the separation medium, whereas the rosettes are found in the pellet.

The working solution of AET contains 1.876 g of 2-aminoethylisothiouronium bromide (Sigma Chemical Co.) in approximately 30 ml of double-distilled H_2O, adjusted to pH 9.0 with 10 M NaOH; the volume is made up to 50 ml with double-distilled H_2O and filtered through a 0.2 µm filter.

To 15 ml of sheep erythrocytes (supplied in Alsevers solution and available from Tissue Culture Services) 8 ml of sterile normal saline (0.9% NaCl) is added. After centrifugation at 600 g for 5 min, the supernatant is removed using a pipet – not by decantation. This should be repeated three times, with the last centrifugation increased to 10 min. The erythrocyte pellet should be resuspended to 50 ml with the prepared AET solution and incubated for 15 min at 37°C. After incubation the erythrocytes are again pelleted at 600 g for 5 min. This washing step should be repeated at least four times until the pellet loses its sticky quality and the supernatant is clear. The pellet is quite difficult to resuspend, but it must be done thoroughly to avoid clumping. After the final wash the erythrocytes are resuspended in complete medium up to 50 ml. AET-treated erythrocytes can be stored for 1 week at 4°C, although it is advisable to wash the cells before using them.

Lymphocytes prepared from a suitable source (see Sections 10.1.1 and 10.1.2) are resuspended to 1×10^7 per ml. Equal volumes of lymphocyte suspension and AET-treated sheep erythrocytes should be mixed together and 5 ml aliquots of this mixture layered on to 5 ml of lymphocyte separation medium (as in Section 10.1.1) in sterile 25 ml Universal tubes. The tubes are allowed to stand for 10 min at room temperature before being centrifuged slowly at 40 g for 10 min; the speed should then be increased to 200 g for a further 30 min. The tubes must then be carefully removed from the centrifuge and the enriched B-lymphocyte cells at the interface harvested, taking as little of the surrounding liquid as possible. The isolated enriched cells should be washed at least twice in serum-free medium (5 min, 400 g) and counted in the usual way. This procedure can be repeated for increased purity of B-lymphocytes, bearing in mind that further purification will result in a lower yield of cells, increased risk of contamination and decreased viability of the cells. To check that the T-lymphocytes have rosetted the pellet should be checked under the microscope. *Figure 10.4* shows typical rosettes.

10.2.2 Enriching for T-lymphocytes

The ability of B-lymphocytes, which have surface immunoglobulins, to attach to nylon wool can be used as a method of negative selection for T-lymphocytes.

The nylon wool (scrubbed nylon fiber available from Travenol) must be carefully prepared by prewashing in 2 M HCl, thoroughly rinsing in

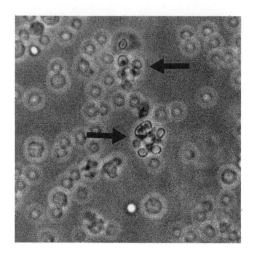

FIGURE 10.4: AET rosetting of human T-lymphocytes. Two rosettes are shown here surrounded by numerous single sheep red blood cells and white blood cells. Visualizing rosettes in focus results in the background cells being out of focus.

double-distilled water (at least five or six times), dried, teased apart with forceps and gently plugged into an autoclavable (i.e. polyethylene not polystyrene) plastic syringe (0.6 g of nylon wool in a 10 ml syringe barrel, 1.2 g in a 20 ml syringe or 0.12 g in a 2 ml syringe). The syringe should be packed to a little more than half its volume. A short length of autoclavable plastic tubing is placed on the end of the syringe and a clamp attached to enable the flow of medium to be regulated (*Figure 10.5* is a diagram of a nylon wool column). Foil can be conveniently used to make a cap to put over the top of the syringe and also to contain the tubing end: this will protect the assembly against contamination when it is moved from the hood to the incubator during the procedure. This assembly should now be placed in an appropriate container and autoclaved.

The lymphocytes to be used should be separated as described in Section

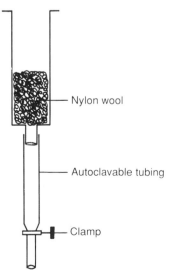

Nylon wool

Autoclavable tubing

Clamp

FIGURE 10.5: A diagram of a syringe filled with nylon wool attached to tubing with a clamp to regulate medium flow.

10.1, counted and resuspended to 5×10^7 per ml. If using a source of lymphocytes containing a relatively high starting concentration of B-lymphocytes (e.g. mouse spleen) a total of 1×10^8 cells can be added to a 20 ml syringe. This number can be increased (up to double) for a human lymphocyte suspension, where the starting concentration is less, i.e. 15–20% B-lymphocytes.

Complete medium (e.g. RPMI-1640 + 10% (v/v) serum) should be warmed to 37°C and the column set up in the cell culture hood with the plastic tubing at the bottom of the syringe clamped. The warmed medium is added to the top of the nylon wool in order to preincubate the prepared columns; once the nylon wool is soaked in medium, the clamp is released. In this way, two syringe volumes of medium should be washed through the column. The last wash volume is run out until it reaches the top of the nylon wool, and the tubing is then clamped.

The cell suspension should be added to the syringe and allowed to run into the nylon wool by briefly releasing the clamp. A volume of medium equal to that of the cell suspension is applied, and again run in to the nylon wool. Then the assembly is incubated in a 37°C incubator for 45–60 min (remembering to replace the protective foil before moving from the hood).

Elution of the unbound lymphocytes is achieved by adding more medium to the top of the column (approximately one syringe volume), and releasing the clamp to allow the medium to run out slowly at a rate of approximately 10 drops/min. The eluted lymphocytes should be washed twice before use by centrifugation at 200 g for 5 min.

10.2.3 Alternative methods of selection

A popular alternative method is the use of magnetic beads ('Dynabeads' from Dynal) coated with a monoclonal antibody to a cell surface antigen. The beads are exposed to a cell suspension; those cells that are attached to the antibody on the beads are pulled out of the suspension by a magnet. The other cell types in the suspension are left as an enriched population. The magnetic bead method involves a certain amount of investment, i.e. magnetic beads, monoclonal antibodies and the magnet. Good technical information, including reference lists, is available from the manufacturers.

Other techniques include variations on the 'panning' technique: this is the adsorption of cell surface antibodies to a plastic dish before applying the cell suspension for selection. Other adaptations of this have produced antibody covalently linked to the inner surface of tissue culture flasks; this is claimed to be superior to the adsorption method and causes more cells to attach to the plastic surface (e.g. MicroCELLector from Applied Immune Sciences).

More advanced column affinity methods are also available for the selection of T-lymphocytes (e.g. T-cell separation kits from Cedarlane). In a similar way to nylon wool columns, B-lymphocytes are bound on to the column matrix and an enriched T-lymphocyte population is found in the eluate.

10.3 Growing lymphocyte cell lines

Lymphocytes may be grown as cell lines by a number of techniques, e.g. immune stimulation, use of specific growth factors, and via viral immortalization. T-lymphocyte lines can be developed by variations on a basic technique, just one of which will be briefly described here. T-lymphocyte lines can be established directly from a basic lymphocyte separation as described in Section 10.1. Growing B-lymphocytes as cell lines has proved more difficult, but Banchereau *et al.* [2] have defined parameters for their long-term growth. A commonly used alternative is to immortalize human B-lymphocytes with Epstein–Barr virus (EBV), a method for which is given in Chapter 13.

10.3.1 T-lymphocyte lines

T-lymphocyte lines can be selected from a lymphocyte source (e.g. human blood or mouse lymph node) following stimulation with antigen or mitogen, and expanded by proliferation in response to interleukin-2 (IL-2, T-cell growth factor) (*Figure 10.6*). Generally, the culture of T-cell lines follows a cyclic pattern of stimulation and expansion, which can include periods of 'rest' for the cultures when no stimulatory factor is present. Variations in methods do exist, for example the incubation period for the stimulation and proliferation of the lymphocytes.

• The antigens against which lines are raised are many and varied, e.g. a whole cell or a soluble protein. In the development of murine T lines the mice used will need prior immunization with the relevant antigen. Generally 10–50 µg of a soluble protein antigen is mixed with complete Freunds adjuvant and injected at the base of the tail; the para-aortic and inguinal lymph nodes are subsequently used as the lymphocyte source. Work with animals requires the appropriate authorization and expertise. References [3–5] give further information concerning immunization and animal handling.

• The *in vitro* development of both human and murine T-lymphocyte lines requires the addition of antigen to the cell suspension. If the antigen concentration to be used *in vitro* is not known, then lymphocytes can be set up with a range of concentrations and the potential lines selected following observation of the growth achieved. However, it is preferable to obtain a more precise definition of the antigen concentration required by a lymphocyte transformation assay. If using mice, then an animal immu-

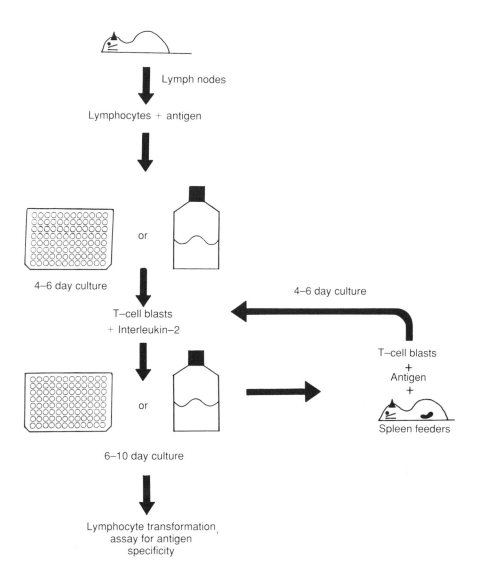

Lymph nodes

Lymphocytes + antigen

or

4–6 day culture

4–6 day culture

T–cell blasts
+ Interleukin–2

T–cell blasts
+
Antigen
+

Spleen feeders

or

6–10 day culture

Lymphocyte transformation
assay for antigen
specificity

FIGURE 10.6: *Selection of T-lymphocytes. Flow diagram illustrating a basic method for creating T-lymphocyte cell lines.*

nized using the same protocol as that used for establishing the line should be used for the transformation assay. If human lines are to be developed, then the same donor should be used. See Section 10.3.2 for further information on this method.

• When developing human T-lymphocyte lines some laboratories use heat-inactivated pooled human serum. Human serum can be filtered by using a step-down system of prefilters – 0.45 μm and 0.2 μm filters with a positive-pressure system for large volumes. Alternatively, small amounts of medium + 5–10% (v/v) human serum can be filtered quite easily through 0.2 μm disposable filters for immediate use.

• If human T lines are being established, then it must be remembered that the donor of the initial T-lymphocytes must supply the feeder cells for subsequent line growth and cloning. Mouse lines must be maintained by feeder cells of the same strain.

The isolated lymphocytes should be adjusted to a concentration of 2 × 10^6/ml in medium plus 10% (v/v) serum (RPMI 1640 is recommended). Antigen should be added to the suspension as appropriate and 1–2 ml placed in 24-well plates. (It is also possible to use 5–10 ml of lymphocyte suspension in 25 cm^2 cell culture flasks placed upright in the incubator.) Due to the high initial cell density, feeder layers are not required at this point. These cultures are incubated for a period of 4–6 days, after which time blast-like cells should be seen when observing the cultures under the phase contrast inverted microscope. The blasts are large, irregularly shaped and phase-bright cells. They can be harvested by centrifugation at 150–200 g for 5 mins and counted.

• Some workers separate the blasts from other cells by density gradient centrifugation, but this should only be attempted if there are a large number of blasts (i.e. approximately 5 × 10^6 or more) because many cells can be lost. Lymphocyte separation medium can be used as before, or a commercial separation medium is available for mouse cell preparations.

The blasts should be adjusted to a concentration of 1–2 × 10^5 cells/ml in medium plus recombinant IL-2 at a concentration of about 5–10 U/ml, and added to 24-well plates or 25 cm^2 upright flasks. The cultures should be incubated for approximately 7 days, although this is another variable factor and the cultures should be checked by observing under the microscope. Feeder cells will not usually be required for this stage as the addition of IL-2 will serve to expand the culture. There are, however, a number of valid alternatives to this method: mouse T-cell lines, for example, are often 'rested' without antigen stimulation or IL-2 at this stage, by simply adding 1–2 × 10^6/ml irradiated feeder cells to the harvested blast suspension (2 × 10^5/ml) and incubating for approximately 7 days. These cultures are then expanded when required by the addition of IL-2.

• There are alternatives to using recombinant IL-2: producer cell lines (e.g. MLA-144, gibbon lymphoma) will produce IL-2 which will suffice for both mouse and human work. The supernatants from these lines should be tested regularly for activity (see Section 10.3.2 for testing on PHA or ConA blasts). In general, approximately 10% (v/v) of supernatant is used but this depends on the strength of the supernatant produced. Collection of supernatant from these cells is simply achieved by growing cells to confluence, centrifuging the cells out of the medium and subsequently filtering the supernatant with a 0.22 μm filter prior to testing the activity. The supernatant should be stored in small frozen aliquots.

After culture with IL-2 (or 'resting' of the culture) the blasts should once again be stimulated with antigen in the presence of irradiated (30 Gy) autologous (i.e. derived from the same source as the original lymphocytes) feeder cells (at $1–2 \times 10^6$/ml of culture).

The cycles of antigen stimulation and rest will need to be continued for growth of these lines. Once lines have been established, it is important to test for their specificity by a lymphocyte transformation test (see Section 10.3.2).

10.3.2 Testing for specificity

Testing for the specificity of a lymphocyte's proliferative response to antigen should be performed at an early stage of cell growth (see *Figure 10.7a*). Briefly, 2×10^5 lymphocytes (i.e. 100 µl of 2×10^6 cells/ml) and 100 µl of various antigen concentrations are added to each well of a 96-well round-bottomed plate in triplicate wells. After 4–5 days 20–40 kBq (0.5–1 µCi) of [^{3}H]-thymidine should be added per well and the plates incubated for 6–16 h before harvesting the cells on to glass fiber filters and counting in a scintillation counter. Remember that appropriate positive and negative controls need to be included. Mitogen should be included for a positive control, whereas medium alone can be used for a negative control, but preferably a non-specific antigen should also be included. Cell proliferation is measured by the rate of incorporation of [^{3}H]-thymidine into DNA. The simplest procedure is to use an automatic cell harvester (e.g. Automash™) to collect each cell sample on a filter and then to lyze the cells by washing with distilled water. This process elutes all residual radiolabel and the [^{3}H]-DNA remains on the filter. In a typical proliferative response the counts in the [^{3}H]-DNA should be greater in the tests than in the controls.

The release of growth factors such as IL-2 can be tested on phytohemagglutinin (PHA) (human) or Concanavalin A (ConA) (mouse)-induced blasts. Once again [^{3}H]-thymidine can be utilized to detect proliferation of the blast cells in response to potential factor release from the T-cell line (see *Figure 10.7b*). Using 2×10^6 cells/ml (e.g. mouse spleen lymphocytes or human peripheral blood lymphocytes), 2 µg/ml ConA or 1 µg/ml PHA are added as appropriate, and left in upright 25 cm^2 flasks for 3 days. Blasts should be washed by centrifugation and resuspension three times, and subsequently used at 2×10^5/ml in 96-well plates for testing with supernatants (each in triplicate). Negative (medium alone) and positive (IL-2) controls in triplicate should be included. The plates are left for 16 h, with the final 4 h including 20–40 kBq (0.5–1 µCi) of [^{3}H]-thymidine.

• For detection of inhibitory factors a known amount of IL-2 is added plus the supernatant to be tested. These supernatants should be left in contact with cells for 16 h, with the last 4 h including 40 kBq (1 µCi)/well of [^{3}H]-thymidine.

(a) Lymphocyte transformation assay

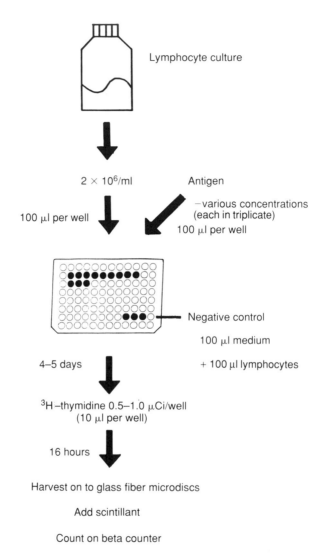

FIGURE 10.7: *Specificity of proliferative response.* (**a**) *A lymphocyte transformation assay to test for the specificity of an antigen response of selected lymphocytes by assessing the incorporation of [³H]-thymidine into DNA.*

• Some factors stimulate or inhibit macrophages: a nitroblue tetrazolium assay can detect this effect on normal mouse macrophages. For more information see reference [6].

10.3.3 Cloning the lines

The cloning of lymphocyte lines can be performed in the same way as shown in Chapter 9, though with the following refinements. Irradiated

(b) Con A/PHA blast assay

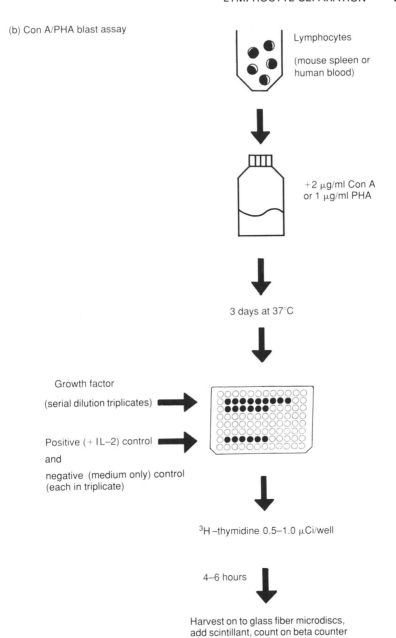

Lymphocytes

(mouse spleen or human blood)

+2 μg/ml Con A or 1 μg/ml PHA

3 days at 37°C

Growth factor

(serial dilution triplicates)

Positive (+ IL–2) control

and

negative (medium only) control (each in triplicate)

³H –thymidine 0.5–1.0 μCi/well

4–6 hours

Harvest on to glass fiber microdiscs, add scintillant, count on beta counter

FIGURE 10.7: *Continued (**b**) Mitogen-stimulated lymphocytes can be used to test for the presence of growth factors (or inhibitors) by assessing the incorporation of [³H]-thymidine into DNA.*

feeder cells (30 Gy) should be added at $1–5 \times 10^5$ cells per well (for 96-well plates where each well contains a total volume of 200 μl) and antigen. Some workers add mitogen to the cloning plates to assist stimulation of the specific lymphocytes, e.g. 1 μg/ml PHA for human T lines).

Wells showing growth should be expanded into 2 ml cultures in 24-well plates, each well containing 1×10^6 irradiated autologous feeder cells, antigen and IL-2. Once clones have been selected by functional tests, then further expansion into small flasks can be followed by resumption of the cycle of antigen with feeders and IL-2 treatment as before.

References

1. Ford, T.C. and Graham, J.M.(1991) *An Introduction to Centrifugation*, BIOS Scientific Publishers, Oxford.

2. Banchereau, J., De Paoli, P., Valle, A., Garcia, E., Rousset, F. (1991) *Science*, **251**, 70.

3. Johnstone, A., Thorpe, R. (1987) *Immunochemistry in Practice*, Blackwell Scientific Publishers, Oxford.

4. Weir, D.M. (ed) (1986) *Handbook of Experimental Immunology* 4th edn, Blackwell Scientific Publishers, Oxford.

5. Harlow, E., Lane, D. (1988) *Antibodies: A Laboratory Manual*, Cold Spring Harbor Laboratories, New York.

6. Dockrell, H., Taverne, J., Lelchuck, R., Depledge, P., Brown, I., Playfair, J. (1985) *Immunology*, **55**, 501.

11 Cell Fusion

11.1 Aims and requirements

The best-known type of cell fusion is that of antibody producing normal murine B-lymphocytes from an immunized mouse with a murine myeloma cell line: this enables a monoclonal antibody of choice to be produced by a permanent hybridoma cell line. Other cell types can also be fused, e.g. T-lymphocytes fused to T-lymphocyte lines and the fusion of mouse and human cells to produce human chromosome libraries. A fusion of mouse and human cells will preferentially eject the human chromosomes in a random manner, leaving just one or part of a human chromosome in each fused cell. Other types of cell fusion enable the dominant/recessive nature of certain cell functions to be determined; a function$^+$ cell can be fused with a function$^-$ cell, e.g. the fusing of tumor cells with non-tumor cells to establish the dominant/recessive nature of the genetic lesion(s). Other examples of cell–cell fusion are all based on the same basic principle of selection of fused cells from the background of parent cells.

The fusion procedure itself is relatively simple: two parent cells are incubated with polyethylene glycol (PEG). PEG is a 'fusogen', which acts by fusing the membranes of adjacent cells. It is not the only fusogen available, but it is the most convenient and commonly used. Following the fusion, selection of the fused cells is most commonly achieved by culturing them in 'HAT' medium. This is complete culture medium (e.g. RPMI 1640 or DMEM) containing hypoxanthine, aminopterin and thymidine. The aminopterin blocks steps in the *de novo* synthesis of nucleotides, whereas hypoxanthine and thymidine are intermediate metabolites used in 'salvage pathways' which can bypass the *de novo* blockage. The myeloma cell line cannot use the salvage pathway because it has been selected to be deficient in the salvage pathway enzyme hypoxanthine guanine phosphoribosyl transferase (HGPRT), therefore these cells will die in HAT medium. A fused cell will remain viable because it will utilize the salvage pathway enzymes inherited from the lymphocyte parent, whereas the mouse lymphocytes will die off naturally. *Figure 11.1* illustrates the strategy behind B-lymphocyte fusion strategy.

Before a fusion can be attempted the mice must be immunized and

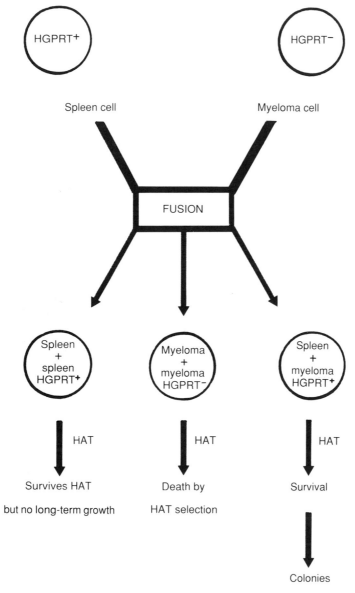

FIGURE 11.1: *The strategy behind the use of HAT medium (hypoxanthine, aminopterin and thymidine) to select for a fused cell which is HGPRT⁺. The enzyme HGPRT ensures that the cell can avoid using the de novo synthesis pathway for nucleotides which is inhibited by aminopterin in HAT medium.*

boosted according to a specific immunization schedule, recommended for the particular antigen to be used. An immunization schedule cannot be hurried and requires planning prior to the fusion taking place, to establish a suitably immunized animal to be ready at the right time. Generally, a priming injection of 10–50 μg of soluble antigen mixed with complete Freund's adjuvant should be given, followed by a boost 2–3 weeks later (usually in Freund's incomplete adjuvant). The spleen cells can be used 3–4 days after the boost. Freund's adjuvant is an oil which, when mixed with antigen, makes an oil/water emulsion which allows slow release of the antigen. Freund's is called 'complete' when heat-killed mycobacterium is included with the oil adjuvant to stimulate the immune system.

In addition to the immunization schedule, a 'screening' procedure for the desired end-product of the fusion must be prepared and ready for use when the hybrid cells start growing. This screening procedure could test for the presence of antibodies via immunofluoresence or enzyme-linked immunosorbent assay (ELISA), or by karyotypic analysis to give the number of chromosomes in a particular hybrid in conjunction with drug selection. The screening procedure can be time-consuming but is an important part of a fusion strategy; however, it is not within the scope of this book to give comprehensive details of immunization and screening procedures. Excellent advice can be found from various sources, and we particularly recommend references [1–3].

The culture medium and serum to be used should be carefully selected. In response to the needs of fusion technology, cell culture suppliers have created ranges of media, sera and supplements specifically for use in fusion procedures, cloning and culturing hybrid cells. It is necessary to obtain a good serum batch, which has been tested and shown to support sensitive cell lines at low densities. Many laboratories use a medium such as RPMI 1640 or DMEM supplemented with a tested serum and find this perfectly adequate. Alternatively, some laboratories now prefer to use a serum-free enriched medium which has a number of defined additional factors (serum replacement factors), e.g. transferrin, insulin and selenium dioxide to name just three (see Chapter 4). Further information should be sought from the technical sections of the tissue culture supplies catalogues from Sigma and Gibco. HAT solution (hypoxanthine, aminopterin and thymidine) can be purchased as a 50 × or 100 × concentrated stock solution from cell culture suppliers.

11.2 Fusion partners

For B- and T-lymphocyte fusions some favored 'partners' have become well established. There are also lesser-used alternatives that have been selected for a specific purpose by a particular laboratory, but it is not possible to list all alternatives here. The common partners for B-lymphocytes have the characteristics of dying in HAT medium and not producing immunoglobulin, which may otherwise interfere with the specific antibody to be produced by the fused cell; examples are X63Ag8.653 [4] and SP2/O-Ag14 [5]. These lines are derived from a mineral oil-induced tumor of a BALB/c mouse, therefore autologous cells used in the overall procedure of producing hybridomas are usually readily available. T-lymphocyte partners include TIB 48, an 8-azaguanine-resistant subline of the AKR thymoma BW5147 [6–8]. BW5147 is available from both the European Collection of Animal Cell Cultures and the American Type Culture Collection (see Appendix B). For ease of hybrid analysis it is preferable to use a non-AKR strain of mouse for the lymphocyte partner, for example strain C3H, which also has the advantage of being histocompatible with (i.e. will not mount an immune response against) the AKR strain and can be used to provide feeder cells.

Where the production of B- or T-lymphocyte hybrids is not the aim, the selection of partners may be made on different criteria. Often both partners need selectable markers because they are cell lines, in contrast to lymphocytes, which will die off naturally after 2 weeks in culture.

As described in Section 11.1, the myeloma parent line must be HGPRT$^-$. This can be maintained, prior to the fusion, by treating the line with 6-thioguanine (6-TG) or 8-azaguanine (8-AG), neither of which is harmful to those cells without the HGPRT enzyme. 6-TG and 8-AG are purine analogs which can be incorporated into DNA via HGPRT through the 'salvage pathway', and cause cell death. However, if cells are devoid of the HGPRT enzyme then the analog will not be incorporated into the DNA because the *de novo* pathway is being utilized, which does not require purines or their analogs, relying on simpler basic molecules. To maintain the cells as HGPRT$^-$ they should be grown in 2×10^{-5} M 6-thioguanine or 20 µg/ml 8-azaguanine. The cells will continue to grow satisfactorily if they retain resistance, although a few cells will die off. However, if all the cells die it is possible that the concentration of the 6-TG is incorrect. After 2–3 days the cells should be subcultured into medium plus 6-TG again. Finally, the cells are subcultured after 2–3 days back into medium + 10% (v/v) serum without 6-TG or 8-AG. The cells should be suitable as a fusion partner once this has been completed. Obviously, if the cells escape resistance (i.e. they are HGPRT$^+$) they are of little use in fusions, as they will not die in HAT medium as required.

Examples of alternative strategies for fusing different cell lines are shown in *Figure 11.2*. Both the nucleotide salvage pathway and drug resistance selection can be utilized.

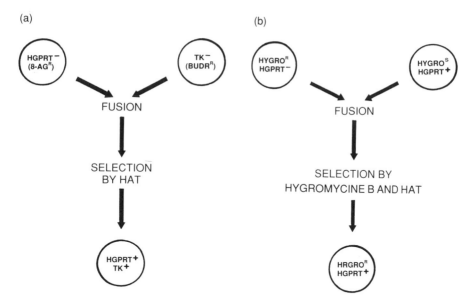

FIGURE 11.2: *Some alternative fusion strategies for cell lines. (a) Cells deficient in hypoxanthine guanine phosphoribosyl transferase (HGPRT⁻), and thus also resistant to 8-azaguanine (8-AGᴿ) can be fused with cells deficient in thymidine kinase (i.e. TK⁻, also conferring resistance to the toxic thymidine analog 5-bromodeoxyuridine, i.e. BUDRᴿ). Selection with HAT medium will kill both parents while allowing fused cells to survive. (b) Cells transfected with the hygro gene (conferring resistance to the effect of hygromycine B) but also deficient in HGPRT (thus sensitive to HAT treatment) can be fused with so called 'wild-type' cells; these are sensitive to the effect of hygromycine B but not HAT treatment (i.e. HGPRT⁺). In this case fused cells are selected on the basis of their resistance to a combination of both hygromycine B and HAT.*

11.3 Fusing the cells

As with many widely used techniques, almost every user alters the basic method in some way to suit him or herself, therefore the method shown here will differ from others in some respects. Because of this, alternative strategies within the basic method will be pointed out where applicable. Hopefully this will encourage some experimentation to find an optimal technique for each application.

11.3.1 Preparation for fusion

If the fusion is to employ mouse lymphocytes, the mice should be boosted 3–4 days before the planned fusion (see Section 11.1).

The fusogen, polyethylene glycol, should be prepared in advance of the day of the fusion. Although most workers use PEG 1500, it is available in different molecular weights from 1000 to 4000, all of which have been used with success in fusions. It is customary to weigh 0.5 g of PEG into glass bijou bottles, which are then sterilized by autoclaving and stored at room temperature until required.

The myeloma (or other partner) cells to be used should be in an exponential growth phase when used for a fusion. Subculturing 24 h prior to the fusion is recommended, while remembering to prepare plenty of cells, as at least 2×10^7 cells will be required after washing by centrifugation.

HAT can be purchased as a $100 \times$ stock solution or made up from the separate components, using 136 mg hypoxanthine, 38 mg thymidine and 1.76 mg of aminopterin in 100 ml cell culture quality water which should be filter sterilized. The HAT stock should be diluted in complete medium (containing 20% (v/v) serum) to give final concentrations of hypoxanthine: 1×10^{-4} M, aminopterin: 4×10^{-7} M, thymidine: 1.6×10^{-5} M.

The number of plates to prepare depends largely on the number of cells used in the fusion. For example, if 1×10^8 lymphocytes are being used from one spleen, together with 2×10^7 myeloma cells, then four plates should be considered. A certain cell density in the wells is thought to be important during the initial days of the fusion, but of course this can be altered based on personal experience of the system and in the light of observed fusion efficiency.

A fusion will produce cell debris from unfused cells dying off, therefore an option is to use feeder layers consisting of peritoneal macrophages which can help to clear debris. These should be prepared the day before the planned fusion and, as a general rule, the macrophages obtained from the peritoneum of one mouse suffice for four 96-well plates (remember that it is recommended that the edge wells should not be used). The macrophages should be plated out in medium containing 20% (v/v) serum + HAT at 100 μl per well. The approximate cell concentration should be 2×10^4 per well. Techniques for extracting peritoneal macrophages can be found in reference [3].

11.3.2 The fusion

● On the day of the fusion all media and sera should be warmed to 37°C and, ideally, a warming plate should be available on which to keep the fusion partners at 37°C during the fusion process. *Figure 11.3* illustrates the steps involved in the fusion procedure.

The parent lymphocytes should be prepared according to Section 10.1. If a spleen is being used as the source of lymphocytes, a possible option is to remove the red cells of the spleen by ammonium chloride lysis (see Section 10.1.3). It appears, however, that the removal of red cells is not a crucial step since many workers report successful fusions without it.

At this stage a decision must be made on the number of parent cells to be added together; some workers add the partners in equal numbers (e.g. 5 ×

10^7 of each cell type) others use a 2:1, 5:1 or 10:1 ratio (lymphocytes:myeloma cells). However, the 5:1 and 10:1 ratios tend to be the most commonly used (e.g. 1×10^8 lymphocytes and 2×10^7 myeloma parent cells). Both fusion partners should be centrifuged and washed into serum-free medium twice before they are counted and placed together in a 30 ml Universal tube at the chosen ratio. This cell mixture should be centrifuged at 200 g for 8 min and the supernatant removed as completely as possible with a pipet.

While the parent cells are being prepared the PEG should be melted by warming and 0.5 ml of serum-free medium added, giving a 50% (v/v) PEG solution. Some workers add 50 µl DMSO to this mixture to further assist

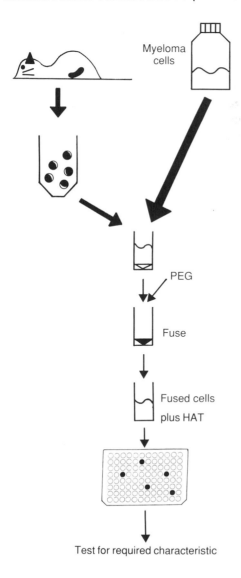

Myeloma cells

PEG

Fuse

Fused cells plus HAT

Test for required characteristic

FIGURE 11.3: A typical fusion procedure, see text for details.

the fusion of membranes of adjacent cells since DMSO tends to make membranes more fluid.

With the tube of cells on the warming plate the pellet should be gently but thoroughly resuspended and the warmed 50% (v/v) PEG added slowly over a period of 1 min. This addition should be performed at 37°C and the tube shaken gently in order to achieve as much contact as possible between the PEG and the cells. 1 ml of serum-free medium should now be added gently and slowly over 1 min, with the tube being shaken gently and constantly; then another 8–9 ml of serum-free medium is added over a period of 5 min. Subsequently the tube should be filled with warmed serum-free medium and centrifuged at 200 g for 8 min. The cells can be susceptible to damage by rough handling at this stage, so care should be taken.

The amount of medium used to resuspend the cell pellet depends on the number of cells used in the fusion. For example, 1×10^8 lymphocytes and 2×10^7 myeloma cells could be spread over a total of four plates, using 60 wells of each 96-well flat-bottomed plate. Therefore, the pellet should be resuspended in 24 ml of warmed medium containing 20% (v/v) serum and single-strength HAT mixture, and then 100 µl added to each well. All wells should have been previously prepared with feeder layers of macrophages in 100 µl of medium plus 20% (v/v) serum and HAT (see Section 11.3.1). The plates are then incubated at 37°C in a gassed incubator.

● Alternatively, some workers recommend the addition of HAT 24 h after the fusion procedure. In this case, 100 µl of fusion mixture are plated out in medium with 20% (v/v) serum, but without HAT. The following day, 100 µl of double-strength HAT in medium/20% serum should be added.

11.3.3 Maintaining the fusion plates

After approximately 4–5 days' incubation, the plates should be replenished with fresh medium by carefully removing 100 µl of medium from the wells and replacing with 100 µl of fresh HAT-containing medium. After 14 days the HAT can be left out of the medium and HT (i.e. hypoxanthine and thymidine without aminopterin) added in its place (this is also available for purchase as a stock solution). This facilitates the gradual dilution of aminopterin from the hybrid cells. HT medium should be used for 7 days before returning the cultures to complete medium alone.

Growth of hybrid cells can take up to 8–14 days to become apparent: once growth becomes obvious it is essential to check fully all plates, well by well, noting the positive ones. It is important to judge the medium changes carefully, as detection of a positive producing well will depend on a good concentration of antibody or growth factor, therefore the medium should not be changed just prior to a planned screening of the hybrid supernatants.

11.4 Screening the hybrid growth

The supernatants should be checked for positivity once they become more acidic (i.e. orange/yellow) and the majority of the positive wells have cells covering about a third of the well. The screening procedure may monitor antibodies, growth factors (e.g. in T-lymphocyte fusions), or functional capacity to respond to a specific antigen.

If the production of antibodies is being tested, then supernatant should be removed from all wells showing growth, and tested by immunofluorescence, ELISA or radioimmunoassay [9]. It is possible to test T-hybridomas initially by checking the phenotype of the cells. The parent line given as an example above (TIB 48) is Thy1.1$^+$ and Thy1.2$^-$; the lymphocytes from other mouse strains used as a partner (i.e. non-AKR but histocompatible) are Thy1.2$^+$. Therefore an actively growing cell with Thy1.2$^+$ may be a hybrid that can be subsequently tested by karyotype analysis to confirm hybrid status. The Gibco catalogue gives a good guide to using Colcemid to perform a simple chromosome count from potential hybrid cells. An example of chromosome number in hybrid T-cells is shown in *Figure 11.4*.

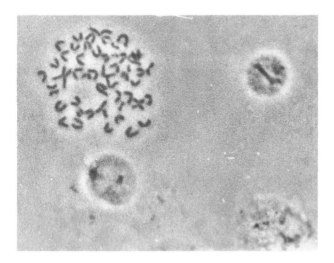

FIGURE 11.4: *One useful way of assessing the success of a fusion is to determine the ploidy of a potential fused cell line. The number of chromosomes from the fused T-cells shown here is greater than the diploid 40 that would be found in an unfused cell. It is rarely exactly double as chromosomes are shed from fused cells quite regularly.*

All negative wells must be discounted quickly to reduce the workload of subculturing non-productive clones. It is beyond the scope of this book to give detailed procedures for all screening strategies, but advice can be found in references [1] and [2]. Where hybrids are selected on the basis of drug resistance the growth of positive cells is self-evident, and therefore the screening procedure is simpler in this respect.

11.5 Expanding and cloning the hybrid

Once the desired hybrid cells have been selected they should be expanded to a point where some vials of cells can be frozen down as a stock, and also cloned as soon as possible.

Positive hybrid cells should be carefully expanded: this can be into multiple wells of a 96-well plate, which are subsequently pooled into 24-well plates, or straight into 24-well plates. These transfers should not be attempted too rapidly as the cell growth may be acutely density-dependent. It is possible to use feeder cells at this stage to avoid density problems with cell transfer, ideally $1–2 \times 10^7$ cells per well of a 24-well plate. When at least one, or preferably two, batches of a particular hybrid cell have been cryopreserved, then they should be cloned. The cryopreserved batches are insurance against contamination or non-growth in the cloning procedure.

Cloning is designed to maintain the positive producer hybrid cell as the dominant cell type. It is common for hybrid cells to lose the ability to produce the positive factor for which they were selected, therefore it is essential that the population should be cloned to reassert the dominance of the producer phenotype. Chapter 9 describes the basic cloning techniques. Either macrophage feeder layers (see Section 11.3.1), irradiated spleen cells (see Section 10.3.1) or thymocytes may be used as feeder cells. Spleen cells should be used at a concentration of approximately 5×10^5 cells/well of a 96-well plate, whereas macrophages should be used at approximately 2×10^4 cells/well.

Cloning should be performed at least twice initially on the hybrid cell selected. If the hybrids are to be cultured on a long-term basis, they should be cloned regularly to maintain the positive phenotype. Should the cells change characteristics or growth patterns then they should be tested and recloned if necessary. It must be stressed again that it is essential to cryopreserve hybrid cells on a regular basis, in order that a good stock can always be called upon should a culturing disaster strike, or if the cells lose their positivity.

References

1. Johnstone, A.P., Thorpe, R. (1987) *Immunochemistry in Practice*, Blackwell Scientific Publications, Oxford.

2. Harlow, E., Lane, D. (1988) *Antibodies: A Laboratory Manual*, Cold Spring Harbor Laboratories, New York.

3. Weir, D.M., (ed) (1986) *Handbook of Experimental Immunology*, 4th edn, Blackwell Scientific Publications, Oxford.

4. Kearney, J.S., Radbruch, A., Liesegang, B., Rajewsky, K. (1979) *J. Immunol.*, **150**, 580.

5. Shulman, M., Wilde, C.D., Kohler, G. (1978) *Nature*, **276**, 269.

6. Kappler, J.W., Skidmore, B., White, J., Marrack, P. (1981) *J. Exp. Med.*, **153**, 1198.

7. Harwell, L., Skidmore, B., Marrack, P., Kappler, J.W. (1980) *J. Exp. Med.*, **152**, 893.

8. Marrack, P. (1982) in Fathman, C.G., Fitch, F.W. (eds) *Isolation, Characterization and Utilization of T-Lymphocyte Clones*, p.508, Academic Press, New York.

9. Billington, D., Jayson, G.G. and Maltby, P.J. (1992) *Radioisotopes*, BIOS Scientific Publishers, Oxford.

12 Cytotoxicity Assays

12.1 Killer cells

Certain subpopulations of lymphocytes have cell-killing (cytotoxic) ability. These cytotoxic cell types include natural killer (NK) cells, lymphokine-activated killer (LAK) cells and cytotoxic T-lymphocytes (CTLs). NK cells are non-T–non-B-lymphocytes that spontaneously lyze foreign cells without prior sensitization (hence the name natural killer cells)[1]. The majority of LAK cell activity appears to be derived from a subpopulation of NK cells bearing the cell surface antigen CD16 in the absence of the cell surface antigen CD3 (so-called CD16$^+$CD3$^-$ cells). Cytotoxic LAK cells are generated from these CD16$^+$CD3$^-$ cells by incubation with the lymphokine IL-2, a polypeptide factor secreted by activated T-lymphocyte cells [2,3]. This lymphokine was originally purified from the supernatant of activated T-cells, but is now widely available as a pure recombinant protein. Once activated, these LAK cells require no prior sensitization to lyze target cells. In this respect they demonstrate the same non-self but non-specific ability to lyze targets as do NK cells; generally this is known as major histocompatibility complex (MHC)-unrestricted cytotoxicity [4,5].

In contrast to both NK and LAK cells, CTLs result from, and require, previous immune stimulation by foreign cells. In other words, CTL precursor cells, when activated by particular target cells, will subsequently be restricted in their lytic ability to those same target cells. Subsequent exposure to these target cells will produce specific (MHC-restricted) cell killing by the cytotoxic T-cells: this can be tested by including a non-specific target cell in the assay [6].

12.2 Requirements of cytotoxicity assays

In *in vitro* studies the cytotoxic lymphocytes (NK, LAK and CTLs) are termed the 'effector' cells, whereas those that are lyzed are called the 'target' cells. The target cell lines are those which can be effectively labeled with radioactive chromium (^{51}Cr) and show only limited spontaneous release of this radiolabel. The chromium, presented to the cells as sodium chromate ($Na_2{}^{51}CrO_4$), is taken up and subsequently oxidized by the cells, the resulting oxide being impermeable to the cell membrane thus preventing the chromium leaking from the cells. Once labeled, the target cells are placed in contact with the effector lymphocytes (at varying concentrations), and only if an effector lyzes a target cell will the ^{51}Cr it contains be liberated. Thus the extent of the lysis achieved is monitored by the removal of the cell supernatants for assay of radioactivity in a gamma counter [7].

• It is vital to stress that ^{51}Cr is a high-energy gamma emitter whose radiation is only stopped by thick lead sheeting. If it is spilt it has the advantage that it can be wiped up and tends not to adhere to a surface in the same way as ^{125}I does. However, very great care should be exercised when using this isotope, and advice should be sought before using it unsupervised [7].

Care must be taken to choose the correct target cell line. For cytotoxic T-cell lines this is predetermined by the type of cells used to sensitize the CTL precursor cells, usually by a mixed lymphocyte reaction. This assay uses approximately $1-5 \times 10^6$ irradiated cells, which are added to the same number of lymphocytes acting as the effector cells, and cultured for 3 days at 37°C.

Not all cell types are susceptible to lysis by NK cells (so-called NK-resistant cells), thus it is important to determine that the chosen target is susceptible. A common line used for the mouse NK system is YAC-1, whereas for the human system K562 cells are often used. K562 cells are a myeloid leukemia cell line with the capability to differentiate into cells which produce hemoglobin. Many other cell lines can also be used as targets for NK cytotoxicity assays – suitability should be assessed by preliminary testing for ^{51}Cr labeling efficiency and lysis susceptibility.

LAK assays are slightly different and need a more careful consideration of the type of target cell to be used. LAK cell precursors are thought to be mainly NK cells. Therefore, after activation to generate LAK cells, by incubation with IL-2 for at least 24 h, a residual number of cells retaining NK activity is to be expected. Thus the targets used should be NK-resistant in order to be sure that the results do not represent a mixture of LAK *and* NK cell killing. Examples of human NK-resistant cells are the human bladder carcinoma line T-24 and the Daudi human EBV-positive Burkitt lymphoma cell line. Once again, some preliminary testing of ^{51}Cr uptake and lysis susceptibility can easily determine other cell types that can be used.

Cytotoxicity assays can be conveniently carried out in 96-well (round-bottomed) plates. The labeled target cells are added to plates in which triplicate samples of effector lymphocytes have been diluted, to produce final effector-to-target ratios of 50:1, 25:1, 12·5:1 and 6·25:1. In effect, a standard number of 1×10^4 target cells is added to each well, with effector cells serially diluted from 5×10^5 cells/well to $6·25 \times 10^4$ cells/well.

12.3 Preparation of the effector cells

The lymphocytes to be used as effectors should be prepared as described in Chapter 10 and adjusted to a concentration of 5×10^6/ml. It is possible to prepare a suspension of lymphocytes and store them overnight at room temperature, in medium that contains Hepes, e.g. BME plus 5% (v/v) serum. If you do this it is advisable to store the cell suspension in a tube (for instance a sterile Universal tube) on its side, so that the cells will not form a compact pellet, which can result in reduced viability of the lymphocytes due to crowding. The suspension should be centrifuged (5 min at 200 *g)* and resuspended in fresh RPMI 1640 + 5% (v/v) serum before use. Lymphocyte effector cells produced from a mixed lymphocyte reaction should be washed by centrifugation following their 3-day incubation with stimulator cells, and resuspended in fresh medium + 5% (v/v) serum.

Prior to addition of the lymphocyte effector cells, a 96-well plate should be prepared by adding medium (RPMI 1640, no serum required) to all the edge wells; 100 µl of RPMI 1640 + 5% (v/v) serum should then be added to the wells indicated in *Figure 12.1*, remembering that each effector-to-target ratio needs to be tested in triplicate. The lymphocytes prepared earlier can now be added in a volume of 200 µl to the indicated first set of triplicate wells (i.e. the 50:1 ratio), to give an initial total number of 1×10^6 lymphocytes within these wells. Using a micropipet, 100 µl of cell suspension is removed from each of these three starting wells (i.e. 5×10^5 cells from each) and added to the next set of triplicate wells (25:1 ratio) and mixed with the medium already present within the well (100 µl) by gently pipeting up and down a number of times. Subsequently 100 µl of the 25:1 dilution wells is removed and added to the next triplicate of wells (12·5:1). This can be repeated as many times as necessary. However, in general a final dilution of 3.125:1 is usually more than sufficient (see *Figure 12.1*). The last set of triplicate wells in the sequence will result in an excess of 100 µl of cell suspension, which should be discarded. All wells should contain 100 µl of cell suspension.

Subsequent manipulation depends on the type of assay being prepared. For an NK or CTL assay, 100 µl of RPMI 1640 + 5% (v/v) serum should be added to all wells being used prior to the addition of labeled target cells. For an LAK assay the medium should contain 1000 U/ml of IL-2 and 100 µl of this is added to give 100 U/well. It is possible to use less IL-2 than this, but a high concentration of IL-2 ensures maximal activation of lymphocytes. It is advisable to include a second plate, identical to the first

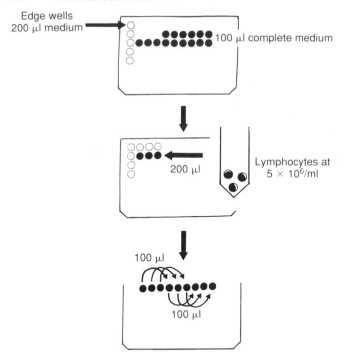

Dilutions to give final effector: target cell concentrations

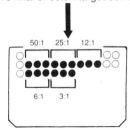

FIGURE 12.1: *The preparation of 'effector' cells in a 96-well plate for a cytotoxicity assay. Simple serial dilutions (in triplicate) taking 100 μl from each set of wells will eventually give a range of concentrations with which to test the target cells.*

in all but the addition of the IL-2; this serves as an important control, demonstrating the absence of LAK activity in the absence of IL-2.

The plates should be incubated for up to 5 days at 37°C (although as little as 24 h incubation may be sufficient). It is our experience that after 6–7 days' incubation the cell-killing ability begins to wane (this may be due to medium exhaustion). We would recommend a particular time period of incubation is chosen (between 1 and 5 days) and adhered to for consistency of experimentation.

12.4 Preparation of the target cell lines

Whatever cell line is being used, a standard method for labeling them with ^{51}Cr, as described here, should be adequate. The target cell line should be healthy and in logarithmic growth, therefore the cells should be checked a number of days in advance of the assay and subcultured as appropriate (see Chapter 5). *Figure 12.2* illustrates the procedure for labeling with ^{51}Cr.

• Great care should be taken when handling the ^{51}Cr-labeled target cells. All operations with ^{51}Cr should be carried out either with the tube in a lead container or behind a lead screen. Any radioactive material should be disposed of in accordance with rules laid down by national agencies and local safety committees [7]. Always consult your radiation safety officer before using any isotopes.

Preparation of target cells

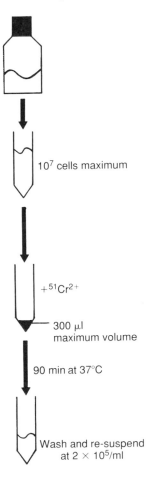

10^7 cells maximum

$+^{51}Cr^{2+}$

300 μl
maximum volume

90 min at 37°C

Wash and re-suspend
at 2 × 10^5/ml

FIGURE 12.2: *The preparation of 'target' cells for a cytotoxicity assay. The resulting ^{51}Cr-labeled cells (at 2 × 10^5/ml) are simply added to the range of concentrations of effector cells prepared as shown in* Figure 12.1.

On the day of the assay the cells should be harvested in centrifuge tubes by centrifugation at 200 g for 5 min, then the cells resuspended in serum-free medium and centrifuged once more. ^{51}Cr-labeling loses efficiency if serum is present, and therefore it should be washed out of the cell suspension. The supernatant should be poured off and the pellet resuspended thoroughly in the small volume of serum-free RPMI 1640 remaining. It is to this small volume of cell suspension that the radioactive chromate is added. The total volume (including the chromate) for the labeling reaction should not exceed 300 μl for efficient uptake.

It is worth noting that commercially available radioactive chromium (as $Na_2{}^{51}CrO_4$) should be diluted into a very small but convenient volume of serum-free medium. The half-life of ^{51}Cr is 28 days, and therefore the volume needed to give 3.7 mBq (100 μCi) will approximately double every month. This amount of ^{51}Cr is sufficient to radiolabel up to 1×10^7 cells. From personal experience we have noted that the type of tube in which the cells are to be labeled can surprisingly affect ^{51}Cr uptake. A relatively large adherent cell type (for instance T24) should be placed in a 25 ml Universal which has a broad base, whereas a smaller cell type such as Daudi should be pelleted in a 10 ml centrifuge tube.

Once the correct volume of ^{51}Cr has been added, the tube should be placed upright in the 37°C incubator for 90 min, with the cap loosened. The tube should be agitated every 30 min during this time to obtain optimum labeling. After incubation the labeled cells should be washed extensively by diluting in 10 ml RPMI 1640 + 5% FCS, pelleting by centrifugation and repeating this three times. After this the cells should be counted with a hemocytometer, resuspended to 2×10^5/ml, and 100 μl of these cells added to each well containing effector cells (see Section 12.5).

12.5 The cytotoxicity assay

- *Figure 12.3* gives a diagrammatic illustration of the procedure.
- For an LAK assay in which the plates have been incubated for more than 24 h, the pellets of effector cells at the bottom of the round well should be resuspended by using a micropipet and mixing by drawing gently up and down once.

It is important to remember the controls that are needed to demonstrate maximum and minimum release of ^{51}Cr from the labeled target cells alone. At least three wells containing 200 μl medium with 5% (v/v) serum will indicate minimal – i.e. background – spontaneous release, and three wells containing 200 μl of 5% (v/v) Triton X-100 will give the maximal release of ^{51}Cr. Triton X-100 is a detergent which will effect the 100% lysis of the 10^4 target cells added to each of the wells.

To add the target cells to the wells of a multiwell plate it is most convenient to use a multidelivery micropipet. Fifty microliters of the 2×10^5

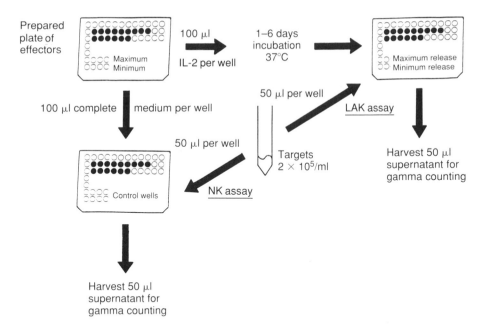

FIGURE 12.3: *Simple protocols for LAK and NK cytotoxicity assays (see also* Figures 12.1 *and* 12.2*).*

cell/ml suspension (i.e. 10^4 target cells) should be added to each well and the plates placed in the incubator for a period of 4 h. Do not leave the plates for longer than this, as the rate of spontaneous release of ^{51}Cr from the target cells increases with time.

After incubation, 50 μl of supernatant from each well should be removed carefully (ensuring that the cell button is not disturbed) into a small plastic tube for counting in a gamma counter. Supernatant removal can be effected either by micropipet or a multichannel pipet, or a more sophisticated removal system such as that available from Skatron (Norway). The samples for counting do not require further treatment but the parameters or 'windows' for counting ^{51}Cr are different from those for other commonly used gamma-emitting radioisotopes, such as ^{125}I. Refer to the appropriate instrument manual or reference [7].

12.6 Results and interpretation

The results from the gamma counter should be analyzed initially by calculating the mean and standard deviation of each triplicate count. The mean counts should be used to work out the percentage killing, i.e. the specific cytotoxicity as shown:

$$\frac{\text{test counts} - \text{background counts}}{\text{maximal counts} - \text{background counts}} \times 100$$

Cytotoxicity can also be expressed in terms of 'lytic units' for human LAK assays, which is defined as the number of effectors required to produce 30% specific cytotoxicity of 10^4 target cells. If the total lymphocyte count is known, then the number of lytic units per liter of blood can be calculated. Additional practical information can be found in reference [8].

References

1. Herberman, R.B., Ortaldo, J.R. (1981) *Science*, **214**, 24.

2. Ortaldo, J.R., Mason, A., Overton, R. (1986) *J. Exp. Med.*, **164**, 1193.

3. Damle, N.K., Doyle, L.V., Bradely, E.C. (1986) *J. Immunol.*, **137**, 2814.

4. Grimm, E.A., Mazumder, A., Zhang, H.Z., Rosenberg, S. A. (1982) *J. Exp. Med.*, **155**, 1823.

5. Grimm, E.A., Ramsay, K.M., Mazumder, A., Wilson, D.J., Djeu, J.Y., Rosenberg, S.A. (1983) *J. Exp. Med.*, **157**, 884.

6. Ucker, D.S. (1987) *Nature*, **327**, 62.

7. Billington, D., Jayson, G.G. and Maltby, P.J. (1992) *Radioisotopes*, BIOS Scientific Publishers, Oxford.

8. Hudson, L., Hay, F.C. (1989) *Practical Immunology*, Blackwell Scientific Publications, Oxford.

13 Epstein–Barr Virus Immortalization

Epstein–Barr virus (EBV) is an extremely useful virus due to its ability to immortalize B-lymphocytes. It is a human herpes-like DNA virus responsible for glandular fever and infectious mononucleosis, and has been implicated in the pathogenesis of nasopharyngeal carcinoma and African Burkitt lymphoma. EBV does not need to integrate into the recipient cell genome to immortalize (although integration does happen). Expression of genes encoded by EBV can (as with SV40, see Section 6) propel a cell out of G_0 and back into the cell cycle. A special feature of this virus is the presence on its genome of sequences that allow its extrachromosomal replication; non-integrated DNA is therefore not lost from the infected cell as it replicates. The restriction of EBV to particular cell types is not due to its immortalizing function but the method of infection. This is mediated by the human complement receptor (known as CR2), which is normally found only on B-lymphocytes and certain antigen-presenting cells. B-lymphocytes can be selectively immortalized, cloned and stored frozen as a permanent viable record that can, at a later date, be recovered for further examination. Additionally these cells are now being used as fusion partners with murine myeloma cells to produce human/mouse hybrids [1].

• Although live virus can be isolated from most healthy people and is generally asymptomatic, work with this virus *must* be performed in a Class II recirculating tissue culture hood. In addition, all plasticware used must be soaked in a strong oxidizing agent such as hypochlorite, before being discarded. If the tissue culture hood has an ultraviolet irradiation unit then it should be routinely used after working with EBV.

13.1 Preparing the virus

There are a number of strains of EBV grown in culture. As with all viruses, EBV is absolutely dependent upon live cells to complete its lifecycle. Some workers collect virus from the sputum of healthy EBV seropositive donors, which is subsequently filtered through a 0.45 μm sterile filter. We would most definitely advise against this, as other viruses may be present and additionally the strain may not be fully characterized. The preferable method is to isolate virus from cultured cell lines. The most often-used strain is an immortalized marmoset 'producer' lymphoblastoid cell line, B95-8, that constantly sheds viable and infectious EBV into the culture medium. The name B95-8 refers more to the strain of the EBV it carries rather than the cell type itself. Avoid a similar cell line called P3-HR1, as this is a viral strain which, although infectious, does not immortalize. The B95-8 cell line can be obtained from the ECACC tissue culture collection (see Appendix B).

Cells are grown in RPMI 1640 + 10% (v/v) FCS and can be frozen down as normal (see Chapter 8). In our experience these cells recover from liquid nitrogen very efficiently and grow with gusto after 24 h in culture. These cells tend to make the medium very yellow (acid) quite rapidly; grow in suspension with a background of adherent cells and tend to clump, making them quite difficult to count. It is possible to disaggregate the clumps by pipeting up and down, gently but thoroughly.

The cells should be thawed and cultured for 24 h (or until viability is >90%) before use. The cells should be subcultured on the basis of keeping them between 2×10^5/ml and 8×10^5/ml until the cell suspension volume equals the volume of virus supernatant required.

• The volume of viral supernatant required should be decided and the B95-8 subcultured until this is achieved. For example, 200 ml of supernatant may be a good idea if a large number of immortalizations are planned, but it will involve freezing 200 aliquots of 1 ml. Otherwise a convenient volume (e.g. 20–30 ml) may be sufficient.

At this point the cells should be cultured until the medium is quite yellow (or cell density is approximately 1×10^6/ml). Overgrowing stresses the cells, causing them to shed virus into a relatively small volume of medium. Additional stressing is possible and there are a number of options:

(1) 24 h treatment with 10 ng/ml of 12-O-tetradecanoylphorbol 13-acetate (TPA).
(2) 48 h incubation at 33°C.
(3) 24 h incubation of cells at room temperature.
(4) Overnight incubation at 4°C.

• We would caution against option (1) as TPA has a number of different effects upon cell lines that are not fully understood, including activation of both T- and B-lymphocytes, and causing many cell types to become adherent and terminally differentiated. Option (2) is probably impractical as few laboratories have an incubator set at 33°C, and frequent resetting of 37°C incubators is not recommended. Either of options (3) or (4) is perfectly adequate, although a theoretical consideration may suggest that the shedding of virus is an energy-dependent process and therefore room-temperature incubation may be preferable. When stressing the cells at room temperature for 24 h the flask is left upright with the cap firmly in place. The following day decant the suspension into a sterile centrifuge tube and centrifuge at 400 g for 10 min.

The supernatant should be harvested and filtered through a 0.45 μm sterile disposable filter to eliminate any remaining cells from the supernatant. Finally, it should be frozen in 1 ml freezing vials, and stored either at −70°C or in liquid nitrogen. Supernatants frozen at −70°C remain viable for at least 1 year.

13.2 Testing the titre of the virus

This is an optional step dependent upon access to lymphocytes from human umbilical cord blood or the ready availability of B-lymphocytes. This test is only essential if undertaking procedures other than merely immortalizing cells, e.g. investigating the use of antiviral agents to inhibit immortalization. EBV will readily infect human umbilical cord blood B-lymphocytes: the T-cells have no prior immunity to the virus and will therefore not interfere with the reaction.

The lymphocytes can be separated in exactly the same way as described for human adult lymphocytes from peripheral blood (see Section 10.1.1). Although there will be nucleated erythrocytes in cord blood, causing the sample/medium interface to appear reddish, this will not interfere with the assay. Nucleated erythrocytes can be distinguished by their small size and characteristic dimple in the middle of the cell. The cells should be resuspended to 2×10^6/ml in RPMI-1640 without serum.

An aliquot of the virus preparation should be thawed and using 500 μl of this virus supernatant, 500 μl RPMI 1640 + 10% (v/v) FCS is added to give a 1:2 dilution. This should be repeated by taking 500 μl of the 1:2 dilution and adding a further 500 μl RPMI 1640 + 10% (v/v) FCS, giving a 1:4 dilution; this serial dilution should be repeated until a final dilution of 1:512 is reached.

Once prepared, 100 μl of each dilution should be added to each of four wells in a microtitre plate (96-well, flat-bottomed). As a control, 100-μl of medium + serum is also added to each of four wells of the same plate, in order that uninfected cell growth and clumping of the EBV-infected cells can be distinguished. Then 100 μl of the prepared lymphocytes (2×10^5 cells) are pipeted into each well containing either viral supernatant or control medium. After 2–4 weeks the wells can be scored for the presence or absence of clumps of immortalized cells, using the non-virus infected control wells as a comparison. The following example shows how the transforming titre is calculated.

Examination of the plates has indicated the following: all wells down to and including the 1:16 dilution contain colonies of infected cells. The 1:32 dilution has three positive wells and one negative. The 1:64 dilution has one positive well and three negative. The 1:128 dilution has only one positive well. All remaining wells, including the control wells, are negative. The four wells at the 1:128 dilution suggest that of the total of 400 µl (i.e. four wells at 100 µl) of virus dilution plated, only one infectious viral particle was present. Thus, the virus concentration in this dilution is 1 per 400 µl, or 2.5 virus particles per ml. As this is the 1:128 dilution, this suggests a viral titre of 2.5 × 128, i.e. 320 virus/ml. The same calculation can be worked through for the other dilutions and will result in different titre estimations, e.g. the 1:32 dilution will give a titre of 240 virus/ml. In general these results should be averaged to achieve an estimate of viral titre. It is recommended that this titre estimate is carried out on more than one sample of B-lymphocytes, as each will have varying susceptibilities to EBV infection; three different samples are suggested for an adequate test.

13.3 Immortalizing adult human lymphocytes

● Cyclosporin A is included in the culture as an inhibitor of T-lymphocyte activation; this prevents the growth of cytotoxic T-cells in the culture that may be directed against the EBV-immortalized B-lymphocytes. It should be dissolved in absolute ethanol to a final concentration of 400 µg/ml and stored foil-wrapped at 4°C (it can be kept for several months under these conditions). If cyclosporin A is not available, preliminary depletion of T-lymphocytes by AET rosetting is recommended (Section 10.2.1).

Human lymphocytes – a total of 1×10^7 is recommended – prepared as previously described (see Section 10.1) are pelleted at 200 g. For every 10^6 lymphocytes used, 20 µl of FCS should be added to the cell pellet. The thawed virus preparation should now be added: normally 1 ml is added and this volume should immortalize many of the B-lymphocytes available. A 'mock infection' performed in the absence of virus can demonstrate the difference between non-specific growth and virus-induced immortal cell growth.

After incubation of the mixture for 30 min at 37°C the lymphocytes should be diluted to 1×10^6/ml (i.e. to 10 ml with 1×10^7 cells) in RPMI 1640 + 5% (v/v) FCS containing 2 µg/ml of cyclosporin A, and plated out in a 96-well plate at 200 µl/well. The plates are placed in the incubator for 1 week, after which 100 µl supernatant should be carefully removed from each well and replaced with 100 µl of fresh RPMI 1640 + 5% (v/v) FCS (without cyclosporin). This should be repeated every 3–4 days.

After 2–4 weeks the growth in the wells can be assessed. Wells showing growth can be subcultured into 25 cm^2 upright flasks containing 3–4 ml of medium + serum, and thereafter expanding the culture carefully. *Figure 13.1* shows the typical appearance of a culture of EBV-immortalized B-lymphocytes. These cells often do not react well to overdilution (especially at an early stage) and dilution to 2–3 × 10^5/ml is therefore recommended. Do not be disturbed if they grow as clumps with a background of adherent cells – this is quite normal, but it can make them difficult to count. Thorough but gentle pipeting up and down will disaggregate them. One of the first aims is to culture enough cells to freeze down a number of vials (see Chapter 8).

13.4 Cloning

The cells which have been isolated from the wells and frozen down are unlikely to be clonal. This is because the original infection and culturing was performed at a high concentration. The cells can be cloned by limiting dilution as described in Chapter 9. Irradiated human lymphocytes (30 Gy) should be used as feeder layers at a concentration of 5 × 10^4/well

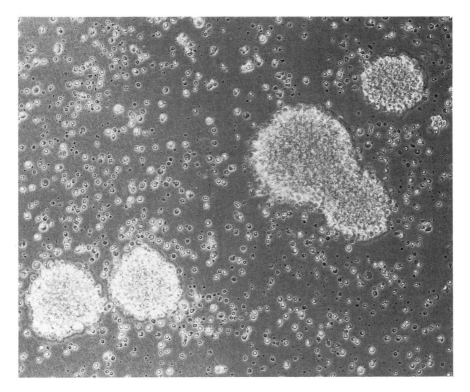

FIGURE 13.1: *A culture of EBV-infected B-lymphocytes. The culture shows typical growth of the cells in large clumps.*

(i.e. 100µl of 5×10^5/ml), to which are added the EBV-immortalized cells at 10, 3 and 0.3 cells/well (see Chapter 9) in the presence of cyclosporin A at 2 µg/ml. A plate containing feeder layers alone in the presence of viral supernatant is a control for the effectiveness of the irradiation. This control ensures that the feeder layers are not themselves immortalized, or if they are it will become apparent.

An accurate cell count is crucial for cloning by limiting dilution, therefore the cells should be resuspended thoroughly. With the suspension in a sterile 10 ml tube the clumps are allowed to settle to the bottom of the tube for 30 min. The top 5 ml of the suspension should be removed and counted. So long as no clumps are present it can be diluted and processed as described in Chapter 9. Otherwise the clumps must be disrupted by pipeting up and down thoroughly, and then the suspension recounted.
• It should be remembered that EBV-immortalized cells may have the ability to shed virus continuously into the medium, and the appropriate safety precautions must be taken when using them.

Reference

1. Harlow, E. and Lane, D. (1988) *Antibodies: A laboratory manual*, Cold Spring Harbor Laboratories, New York.

14 Transfection

14.1 Principles and aims

Transfection is the introduction of foreign DNA into recipient cells. A transfected cell represents the combined efforts of disciplines ranging from molecular biology, microbiology and cell biology to cell culture. Because transfection uses techniques described in this book, and because of its importance in genetic engineering, we consider it an appropriate final chapter and a useful bridge between cell culture and molecular biology for cell biologists.

The bulk of the chapter will be concerned with the tissue culture aspects of transfection, will include some sample procedures for the most commonly used methods, and introduce a range of other approaches.

We do not intend to describe in detail all the background work required for transfecting cells, but we need to give a brief outline of the mechanisms concerned in the provision of DNA in a form that can be used for transfection. Terms that may be used and could confuse those not familiar with the concepts and techniques will be introduced and explained.

14.1.1 Source of the DNA

DNA can be presented for transfection in many different forms, including cDNA clones, genomic DNA, genomic DNA clones and retroviral vectors. What do all these terms mean?

Clone is a term that we have used frequently in this book, and in cell culture it refers to the propagation of identical daughter cells from a single parent cell. In transfection the majority of DNA used is a *molecular clone* of some type. A *molecular clone* is a replica of a specific stretch of DNA or RNA found in a cell. A *cDNA clone* is a piece of DNA that was originally synthesized *in vitro* using cellular RNA as a template. cDNA cloning usually utilizes the viral enzyme *reverse transcriptase*, which uses the information encoded in the RNA to make a mirror image (or complementary) copy of this RNA as DNA (hence the name cDNA or complementary DNA). The RNA is allowed to degrade and the first strand of the cDNA used as a template for a DNA polymerase enzyme to synthesize the second strand of DNA. This DNA is complementary to the first strand of the cDNA and thus identical to the original template RNA. *Figure 14.1* compares the base composition of the original DNA template with its transcribed RNA and the cDNA synthesized from the RNA template.

This double-stranded cDNA must now be preserved in such a way that it can be separated from other different cDNA molecules, stored, amplified, purified and finally used. This is achieved by first *ligating* the cDNA into a *vector*; this construct is now used to transform a bacterial host. Ligation uses an enzyme called *ligase* to form covalent bonds between two pieces of DNA, making a single entity (see *Figure 14.2*). A *vector* is a piece of DNA

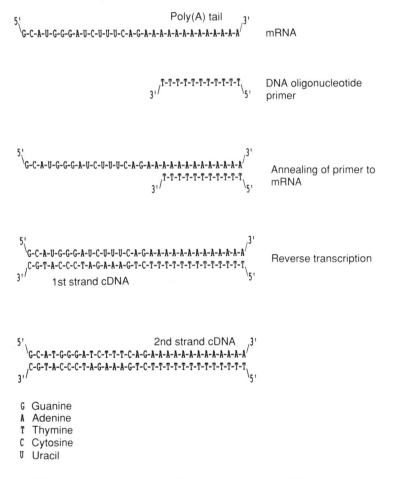

G Guanine
A Adenine
T Thymine
C Cytosine
U Uracil

FIGURE 14.1: *Synthesis of cDNA from poly (A)⁺ RNA using a poly (T) oligonucleotide primer. Compare the base composition of the mRNA, 1st strand cDNA and 2nd strand cDNA.*

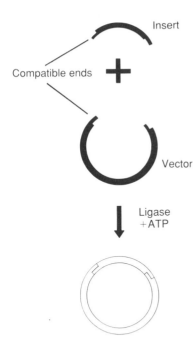

FIGURE 14.2: *Ligation of a restriction digested DNA fragment into vector DNA. The plasmid and insert are digested to give compatible ends, and in the presence of ATP and DNA ligase the fragments become a single molecule.*

that has the sequences required for it to replicate in host bacteria. Thus ligating the cDNA into a vector allows it to be introduced into and grown inside a bacterial cell (*transformation* of bacteria). Vectors come in all shapes and sizes, and there are three basic types: bacteriophage vectors, retroviral vectors and plasmid vectors (*Figure 14.3*).

Bacteriophage vectors are modified from the naturally occurring bacteriophages that infect bacteria. These phages can be engineered (by ligation) to include either cDNA or short pieces of genomic DNA, in addition to their viral DNA. When these bacteriophage viruses infect bacteria, the enzymes in the bacteria are utilized by the virus to make copies of themselves, package themselves in a protein coat, and burst the host bacterial cells and infect other bacteria. Using this system the DNA cloned into the virus can be amplified many times. Indeed, it is quite possible to amplify the number of viral particles in this way by a factor of 10 000 in a matter of hours. These vectors are most often used to plate out and screen cDNA or genomic libraries. A cDNA library represents the RNA content of a culture, tissue or organ (after cDNA synthesis and ligation into a vector). A genomic library, on the other hand, represents the DNA content of the cell, tissue or organ etc. These vectors are generally not used in this form for transfection, but rather to identify the single clone of interest, like taking home a single book from your local library.

Retroviral vectors are modified (genetically engineered) from naturally occurring eukaryote RNA viruses (retroviruses). They can be propagated in bacteria as plasmids, have foreign DNA ligated into them, and be

(a) Bacteriophage vector

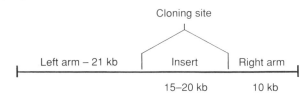

(b) Retroviral vector

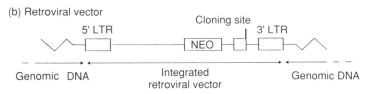

(c) Plasmid vector

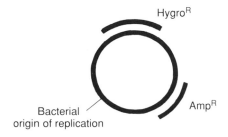

FIGURE 14.3: *Three different types of vector. (a) Bacteriophage vectors of the type depicted can be used for the construction of cDNA and genomic DNA libraries. Some of these vectors can accommodate up to 20 kb of insert by ligation into the cloning site, which can then be propagated in bacteria. (b) Retroviral vectors (depicted here as integrated into the genome of a host cell) can receive genes in the cloning site, and be used to infect and integrate into the genome of the target cell. They can be selected for on the basis of resistance to selective drugs (e.g. the neo gene confers resistance to the antibiotic G418). (c) Plasmid vectors are loops of DNA containing a bacterial origin of replication, a bacterial drug resistance gene (ampicillin), and a multiple cloning site to allow the ligation of inserts. Sometimes a plasmid will also include a gene that will confer drug resistance in eukaryotes (e.g. to hygromycine B in cells in culture). This can be used to select for transfected cells.*

transfected into tissue culture cells. In particular cells called 'packaging cell lines', these retroviruses can return to using RNA as their genetic material, be packaged into a protein coat, shed into the tissue culture medium and be used to infect other cells.

Plasmid vectors are circles of DNA that contain the sequences required for their propagation and selection in a bacterial host as extrachromosomal DNA. Any DNA can thus be ligated into a plasmid and amplified in a bacterial culture.

14.1.2 How can foreign DNA replicate in a cultured cell?

DNA, when transfected into a cell, must be capable of replication in the new host. This is required to prevent the dilution of the small amount of transfected DNA in progressively larger numbers of cultured cells. There are two ways in which replication is achieved. First, the introduced DNA can integrate into the genomic DNA of the host cells (once again by ligation, although in this case the host cell ligase is utilized), and when the cells synthesize new DNA prior to cell division the foreign DNA is replicated as well. Secondly, there are now available special plasmids that contain both the sequences required for independent replication within a bacterial host *and* the sequences required for their extrachromosomal replication in some types of cultured cell lines. These plasmids do not need to integrate into the genome of cultured cells to replicate and, provided that they replicate at least at the same frequency as the host cell, they will not be lost.

There are other cases where the transfected plasmid does not need to remain in the cells for long periods of time: these are called transient transfections. In these short-term experiments the DNA in the plasmid is transcribed into messenger RNA and translated into protein without integration into the host cell genome.

14.1.3 Identification of transfected cells

The first objective of transfection is to overcome the obstruction presented by the recipient cell plasma membrane, thus allowing the foreign DNA to be exposed to the host cell's transcription and replication machinery. This is probably the most efficient part of the transfection process. The next phase is much less efficient: the DNA must be stably integrated into the host cell genome, although for transient transfection this need not matter. This is a very inefficient process which can happen as rarely as one in every 10^6 cells into which DNA has been introduced. A major problem is to identify the cells where DNA has been successfully integrated, and there are a number of ways in which this can be done. Plasmids can be engineered to contain drug-resistance genes in addition to the other foreign DNA that they contain. Thus a cell with this gene integrated into its genome and correctly expressed as a protein product can be resistant to drugs that would normally kill the untransfected cells, and can therefore be selected. Retroviral vectors are usually constructed in this way to contain a drug resistance gene and the DNA of interest.

An alternative to this system is termed co-transfection, where two plasmids are introduced into the cell, one containing the DNA of interest and the other the drug-resistance gene. If the plasmids are mixed in a ratio where the drug-resistance gene is in the minority (often 10:1), drug-resistant cells are usually found to have integrated copies of both plasmids, hence the DNA of interest can be selected.

One special case involves the transfection of one type of cell with the genomic DNA of another; here a cell with integrated foreign DNA will need to display a new phenotype that can be identified. This may be a new gene product that can be assayed, or a transformed phenotype such as growth in low serum concentrations or anchorage independence in soft agar.

14.1.4 Use of transfected cells

There is a multitude of reasons why one would wish to transfect a tissue culture cell, for example, to express a protein encoded by a gene that is not normally expressed in the target cell, thereby enabling a study to be made of the effect of the protein product of this gene on the phenotype of the cells. This type of work is often done with mutated oncogenes, in order to determine the effect of different mutations on the oncogenic or malignant transforming ability of the gene.

Expression of the protein product of a gene at far higher concentrations than normally found in the cell is occasionally required. Using this strategy the protein can then be more easily purified than from the original source tissue.

A common reason for transfecting cells is to facilitate the study of controlling sequences within genomic DNA, and their effect on the expression of reporter genes. In these cases a genomic clone of DNA is ligated at the 5′ end of the reporter gene (upstream), such as the bacterial chloramphenicol acetyltransferase (CAT) gene. The protein product of this gene can be assayed simply, and the expression (i.e. transcription of RNA from DNA) of the gene can be examined under different tissue culture conditions and with different 5′ controlling sequences. *Figure 14.4* illustrates an experiment where different lengths of a genomic sequence have been inserted 5′ to the CAT reporter gene in order to study the effect on transcription (as assayed by CAT activity) of the indicated control regions.

14.2 Transfection of adherent cells

14.2.1 Choice of technique and materials

The first description of the transfection of cells with foreign DNA other than by intact eukaryotic viruses was reported in 1973 [1], using what has become known as calcium phosphate co-precipitation. This was one of the earliest techniques and has undergone extensive modifications, but it is still the method of choice for transfection of most adherent cells. Therefore it is advisable that this should be the first method attempted, even if the cells are only semi-adherent, for example COLO 320 cells.

Transfections can fail to work for many reasons; for instance, the plasmid

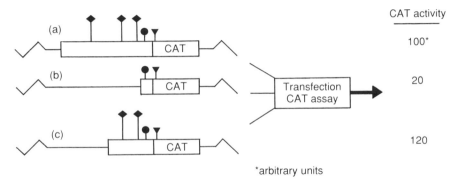

FIGURE 14.4: *Assaying control sequences in genomic DNA. (a) Shows the CAT activity detected in cells transfected with the indicated plasmid; in (b) and (c) parts of the genomic DNA have been removed. The triangle indicates the site of the start of protein synthesis (translation), the circle indicates the site of initiation of RNA synthesis and the diamonds show sites of possible control sequences. After transfection, transcription and CAT synthesis, the activity of the enzyme is assayed. Taking (a) as resulting in a certain level of CAT activity (called 100%), then (b) is reduced to 20%, while (c) is elevated to 120%. These results suggest that the two control sequences in (c) but not in (b) can elevate CAT gene transcription, while the additional control sequence in (a) but not in (c) can repress transcription.*

construct may not be correctly expressed in the target cells. This is difficult to determine empirically, but it is wise to check that the promoter–enhancer sequences in the plasmid have been reported to be active in the cells to be transfected.

The DNA used for transfection must be free from contaminants such as protein or nucleases, and should be supercoiled. Although many workers find that DNA prepared by PEG (polyethylene glycol) precipitation is effective, others claim that one and sometimes two rounds of caesium chloride centrifugation are required to prepare pure protein-free supercoiled plasmid DNA for truly efficient transfection. For further details on the use of CsCl gradients, see reference [2].

If the culture consistently becomes contaminated, try a mock transfection without the DNA. The DNA may not have been originally prepared with transfection in mind: for molecular biology purposes this may not matter, but for cell culture it is crucial.

14.2.2 Calcium phosphate co-precipitation

There are many transfection protocols published: for a detailed review of a number of techniques, refer to reference [3]. Here a generalized technique modified from a number of sources is presented [1, 3–5].
• A number of buffers need to be prepared in advance. The stock buffer used is 2 × Hepes buffered saline (HBS) comprising 1.5 mM Na_2HPO_4, 10 mM KCl, 280 mM NaCl, 12 mM glucose and 50 mM Hepes pH 7.05. This can be made up in 50–100 ml volumes and

filtered through a 0.2 µm filter, discarding the first 10 ml. This is a way of prewashing the filter that also prevents possible pH changes, which could otherwise arise from the buffer's contact with a fresh filter. The filtered buffer can be stored in 2 ml aliquots at $-20°C$ in sterile plastic bijou tubes. Many workers make up a number of $2 \times$ HBS buffers with pH values ranging from 6.95 to 7.1; aliquots from these batches are then tested for the most efficient transfection. The other stock solution is 2 M $CaCl_2$ made up in distilled water, filtered and stored in 1 ml aliquots at $-20°C$.

Two days before transfection the cells should be subcultured into 90 mm tissue culture dishes in complete medium at a concentration of 1×10^5 per dish. This subculturing ensures that after a further 48 h the cells are healthy, in logarithmic growth, and still have room for an additional 24 h growth following the transfection. This assumes that the cells have a doubling time of 24 h; if this is not the case then the plating of the cells should be modified accordingly. On the day of transfection the cells should be examined under the inverted microscope to ensure that they are subconfluent, but could be expected to approach confluence within 24 h. *Figure 14.5* shows fibroblasts ready for transfection.

For each plate to be transfected, 5–30 µg of plasmid DNA, prepared under sterile conditions, are typically required. The DNA should be diluted in sterile distilled water (0.2 µm filter) to give a final volume of 875 µl in a 25 ml sterile Universal tube. With a micropipet, 125 µl of 2M $CaCl_2$ are underlaid gently to the bottom of the tube, followed by 1 ml of $2 \times$ HBS, once again carefully underlaid to the bottom of the tube. The tube can now be swirled gently until all the components are mixed; this will have been completed when the 'streaming' effects observable when one solution mixes with another are no longer seen. Leaving the tube at room

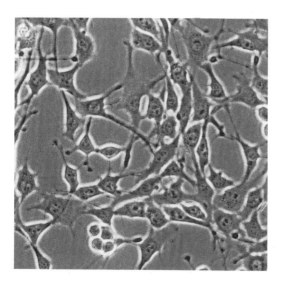

FIGURE 14.5: NIH 3T3 cells ready for transfection 48 h after plating 1×10^5 in a 90 mm plate. The cells are present in sufficiently high concentration to provide target cells, and there is also plenty of room to replicate. Depending on the cell type and growth rate, more cells could be plated if necessary. Opinions are often divided on exactly what the ideal concentration is for transfecting adherent cells.

temperature for 20–30 min allows the DNA–calcium phosphate coprecipitate time to form.

• Many workers find that, when using small amounts of plasmid DNA, the addition of 'carrier' DNA helps maintain transfection efficiency. This can be either commercially available salmon sperm DNA (or its equivalent) or prepared from cells in the laboratory. Used as a sterile solution after being sheared by passage through a narrow-gauge needle, the carrier DNA can be added to the plasmid DNA sample to keep the total content at 30 µg of DNA.

The ideal DNA–calcium phosphate precipitate is so fine that it resembles dilute milk rather than a precipitate, and this should not sink to the bottom of the tube. In practice the precipitates formed can be very variable, and many different methods have been used to ensure their consistency. A crucial factor appears to be the speed at which the precipitate forms, the slower-forming precipitate being the finer and the most effective for transfection. Alternative mixing techniques are used in different laboratories: for instance, the DNA can be gently mixed with the $CaCl_2$ to a final concentration of 250 mM. This is then added dropwise to an equal volume of the $2 \times$ HBS solution while gently swirling the receiving tube. The solutions can also be mixed by gently blowing bubbles into the bottom of the tube with a sterile glass Pasteur pipet attached to an automatic pipet filler.

Once the precipitate has been prepared, the culture medium should be removed from the tissue culture dish. There is no need to wash the cells and the small residue of medium remaining in the plate will not interfere with the transfection [1]. The DNA–calcium phosphate coprecipitate can now be mixed once with a pipet, and placed in the middle of the dish. Rocking the plate gently a few times will ensure that the precipitate comes into contact with all the cells; the lid can now be replaced and the dish incubated at room temperature. After 20 min, 10 ml of complete medium are added gently to the culture (there is no need to mix the added medium) and the plate is placed in the 37°C incubator for a further 3 h. Following the incubation, the medium should be removed, the cells washed once with 10 ml of complete growth medium and finally 10 ml of complete growth medium is added, and the cells are incubated overnight at 37°C.

• Once again there are alternative favored methods. For example, sometimes the cells are not washed after the initial 3 h incubation and the precipitate is allowed to remain with the cells overnight at 37°C. Another method does not involve removing the growth medium from the cells: in this case the precipitate is added directly to the culture and mixed gently with the resident medium [3]. The dish is then returned to the 37°C incubator for 16–24 h, after which the cells are washed with complete medium.

• Transfection efficiency can often be increased with the addition of a glycerol or DMSO 'shock'. When the medium is removed from the cells after the 3 h of incubation, it is replaced with 10 ml of serum-free medium containing either 15% (v/v) glycerol or 10% (v/v) DMSO, and incubated for 2–3 min at room temperature. This is then removed and the cells washed with serum-free medium and cultured in complete medium. Some cells are acutely sensitive to these treatments, therefore this 'shock' technique should be tested beforehand. Comparing the survival of cells 24 h after a shock with untreated cells allows the sensitivity of the target cells to be assessed. A death rate of 10–20% is acceptable; death rates of up to 50% can also be acceptable if the surviving cells are more efficiently transfected. Both the concentration of reagents and the exposure time can be adjusted to suit the cell type.

On the following day the culture should be washed with serum-free medium, trypsinized as normal (see Section 5.4.3) and the cells from one dish divided into five fresh dishes in 10 ml of complete growth medium. After a further 24–48 h of incubation the drug selection technique can be applied for the isolation of stable transfectants, or alternatively the cells harvested for analysis of the expression of the introduced gene. The drug used for selection will depend upon the plasmid used for the transfection, and could include either geneticin G418, hygromycine B, or mycophenolic acid. Whichever is used, a dose–response curve of the drug should be carried out beforehand with the target cell type, in order to determine the concentration at which all untransfected cells die. This is the minimum concentration of drug that should be used.

14.3 Retroviral vector-mediated transfection of non-adherent cells

Suspension cells have for a long time presented problems to those wishing to transfect them, and opinion is divided as to the reasons for this. The problem has stimulated the investigation of a number of alternative strategies which have been successfully used in *some* laboratories, with *some* suspension cells. Calcium phosphate coprecipitation can be used to transfect suspension cells (in much the same way as for adherent cells) but has been found not to work with many cell types and to be inefficient with others. Probably the easiest technique to use – although unfortunately with the most limitations – is retroviral infection. This is also known as retroviral vector-mediated gene transfer.

14.3.1 Characteristics of retroviruses

Retroviruses are viruses whose genetic material consists of RNA. Once the virus has infected a cell, the RNA is reverse transcribed to create a complementary DNA copy. This single-stranded DNA is used as a template for the synthesis of the second strand of DNA, and the double-stranded DNA is integrated into the genome of the infected cell. The integrated retroviral DNA replicates and transcribes itself using the host cell enzymes. The resynthesized RNA is then packaged into a protein coat and shed from the cell as infective virus. *Figure 14.6* illustrates a generalized retroviral lifecycle. Retroviral vector-mediated gene transfer takes advantage of the viral capability to infect and integrate into the host cell genome while disabling viral ability to replicate in the infected cell. This has been achieved by the removal from 'wild-type' retroviruses of crucial genes to create disabled replication-deficient retroviral vectors. These vectors are able to infect cells and integrate the DNA synthesized from the RNA genome template into the host cell DNA. The absence of the genes coding for the protein coat and the reverse transcriptase prevent the replication of the retroviral vector into new infectious viral particles.

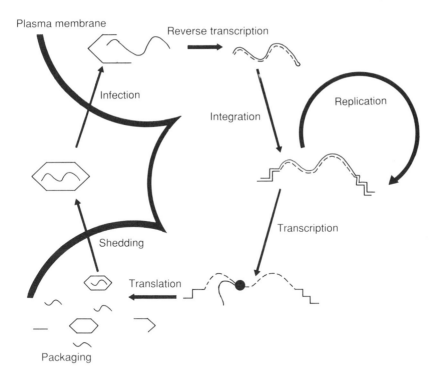

FIGURE 14.6: *Generalized retroviral lifecycle. On infection the RNA genome is released from the protein coat. Reverse transcription results in a RNA:DNA hybrid. The DNA then acts as template for the second DNA strand synthesis, which can then integrate into the genome of the target (or host) cell. Once integrated the virus is replicated along with the host cell. The viral genome can also be transcribed, the RNA translated into protein which is then used to package viral RNA prior to the infectious viral particles being shed from the cell.*

14.3.2 Construction of vectors

The major problem with retroviral infection is not transfection of the target cells, but the availability of the construct containing the gene of interest. The vectors available already have ligated into them drug-resistance genes. For example, the *neo* gene confers resistance to geneticin G418 by coding for the enzyme neomycin phosphotransferase [6] and the *hygro* gene codes for the enzyme hygromycin B phosphotransferase, thus conferring resistance to hygromycine B [7]. These genes allow infected cells to be selected on the basis of continued growth in the presence of selective drugs. The construction of these vectors requires that the gene of interest be ligated into the retrovirus during the DNA plasmid stage of its lifecycle. This construct is then transfected into a specialized cell line (packaging cells) containing a 'helper virus'. The helper virus replaces those functions that have been removed from the retroviral vector and allows it to replicate and be packaged in a protein coat, before being shed into the tissue culture medium as infective retroviral vector particles. These retroviral vector particles are then used to infect another packaging cell line which is then cloned and expanded, and can be used as the source of retroviral vector (*Figure 14.7*). In order to avoid confusion, the resultant clonal cells are referred to as 'producer' cell lines. The infective disabled retroviral vector particles shed by producer cells can then infect and integrate into the genome of the target cell.

14.3.3 Choice of technique and materials

Most producer cell lines are mouse-derived and thus contain mouse helper viruses. Although they are termed amphotropic (meaning they are able to infect across the species barrier), in our experience they are a great deal more effective against mouse target cells than human ones, probably because the protein coat is determined by the mouse helper virus. This is worth remembering when using this system as a transfection technique.

If the correct cell line is available (i.e. shedding a virus containing the correct gene of interest) the infection of the target cells is relatively simple. There are two ways of using a retrovirus to infect target cells, either by co-culture of the target cells with the producer cell line, or by harvesting the virus shed into the medium from the producer cell line. This is then added to the target cell in culture. The latter system is recommended – it is a cleaner culture system and there is less risk of inappropriate cells growing in the culture and interfering with the selection and cloning of transfected cells. Additionally, some cell lines are badly affected by contact with the producer cells and may die or differentiate.

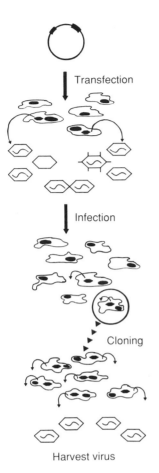

Transfection

Infection

Cloning

Harvest virus

FIGURE 14.7: *Preparation of retroviral vector producer cell lines. Packaging cells are transfected with a retroviral plasmid. The plasmid integrates (by an inefficient process) and these cells produce retroviral particles with the help of an endogenous helper virus. Because transfection can mutate (or otherwise alter) plasmids, some of the retroviral vectors will be defective (as indicated). This mixture of competent and defective retroviral vectors is used to infect packaging cells. The infected cells are selected on the basis of retrovirally acquired drug resistance (indicating the presence of integrated vector), cloned, and the retroviral vector they shed tested for infectivity. In this way the producer cells are known to produce identical and infective retroviral particles.*

14.3.4 Harvesting producer cell supernatant

Disabled retroviral vectors can infect human cells, hence they must be treated as infectious agents and dealt with in the same way as Epstein–Barr virus (see Chapter 13).

Producer cells are adherent cells derived from mouse fibroblasts. When they are growing in culture, virus is shed into the medium, which can be removed from the plate and replaced with fresh medium. This harvested supernatant should be frozen immediately at −20°C. After a further 2–3 h culture, medium can again be harvested. When enough supernatant has been collected the frozen virus can be thawed on ice, pooled, centrifuged at 800 *g* for 5 min to remove cellular debris, and the supernatant carefully removed to a separate vessel. This should be filtered through a 0.45 μm sterile filter, and frozen in small aliquots (2–3 ml) at −20°C. These retroviral vector aliquots are free from the presence of producer cells and can be stored frozen for at least 6 months, but once thawed they should not be refrozen.

At 37°C retroviral vectors have an infectability half-life of approximately

4 h, thus when producer cell supernatant is harvested the only infective virus it contains will have been shed into the medium in the last few hours. This means that the supernatant removed from the producer cell culture every 2–3 h will contain as much infective virus as that taken after 24 h. Thus a single dish of producer cells, when harvested every 3 h, can yield quite large quantities of virus in a single day.

14.3.5 Infection of target cells

The number of cells that need to be infected is dependent firstly on whether they are mouse or human (some mouse cell lines can be infected with close to 100% efficiency), and secondly on whether or not they can be cloned in soft agar.

We will here describe a generalized procedure for the infection of a human cell line that infects with intermediate efficiency (say 3%) and clones in soft agar with an efficiency of 40–50%. Such a cell line is K562 (see Chapter 9). The cells should be used when in the logarithmic phase of growth, the cell concentration adjusted to 2×10^5/ml and 2 ml placed in a 25 cm^2 flask.

Producer cell supernatant (2 ml) should be thawed to room temperature and 32 µl of DMSO and 24 µl of 1 mg/ml Polybrene (filtered through a 0·2 µm filter and stored at 4°C) added. The producer cell supernatant should then be added to the cell suspension and the flask placed upright in the incubator.
• The DMSO and Polybrene are now at a final concentration of 0.8% and 6 µg/ml respectively. These agents have been found to increase the efficiency of retroviral infection in some cases, but may be omitted if the cells are found to be sensitive to their temporary presence. Polybrene is a long-chain polycation that is thought to adhere to both viral and cellular protein, and so facilitate infection.

After 2 h incubation a further 2 ml of producer cell supernatant should be thawed and supplemented with 16 µl of DMSO and 12 µl of 1 mg/ml Polybrene (half the previous amount because the cells are already in the presence of these two reagents). The flask should be carefully removed from the incubator and 3 ml of the supernatant removed (taking care to avoid the cells that will have settled to the bottom of the flask) and replaced with the fresh 2 ml aliquot of producer cell supernatant. This should then be gently mixed with the cells before returning the flask to the incubator as before.
• This procedure can be repeated at 2-hourly intervals for as many times as required (which could, if necessary, include an overnight incubation under the conditions described above). In practice five to six infection cycles should be sufficient.

When the infection regimen has been completed, the cells can be centrifuged for 5 min at 200 g and resuspended in RPMI 1640 + 20% (v/v) FCS before being cloned by limiting dilution or in soft agar in the presence of geneticin G418 (most retroviral vectors confer resistance to G418). If the cells can grow in soft agar this would be the method of choice (see Chapter 9).

• The minimum concentration of G418 (or any other selective drug) that should be used is dependent upon the cell type. However, we would suggest that an *effective* concentration is not less than 300 μg/ml (most G418 purchased is only 50% pure – this should be checked with the manufacturer). Each cell line must be checked, both in long-term growth and in soft-agar cloning for drug sensitivity. It is important that the drug selection is not applied until at least 24 h after the infection.

• When selecting for transfected cells by drug resistance, it is crucial to include an additional set of cultures that have not been transfected but are exposed to the selective drug in the same way. *Table 14.1* shows the results of a typical retroviral vector infection of K562 cells with a *neo*-containing mouse retroviral vector.

14.4 Alternative transfection methods

Space does not permit any detailed recommendations for the numerous techniques used for transfecting cells, but there are a number of alternative methods that may be successful and these are summarized here.

The DEAE–dextran method has its origins in attempts by virologists to increase the infectivity of viruses in cells. They discovered that, whereas the infectivity of some intact viruses was decreased by DEAE–dextran, the infectivity of naked viral DNA (transfection) was considerably increased. DEAE-dextran (another long-chain polycation) appears to bind to both DNA and cell membranes, thus the DNA is closely associated with the cell when it is osmotically shocked. The original method [8] has been variously modified (e.g. see [9]). Most recently, for example, the

TABLE 14.1: Soft-agar selection of retrovirally infected K562 cells

	Colonies per plate Number of cells plated per dish		
	10^4	10^3	10^2
K562 control	*	526 ± 21	58 ± 3.9
K562 + 500 μg/ml G418	0	0	0
KRV control	*	440 ± 66	52.6 ± 5.3
KRV + 500 μg/ml G418	128 ± 9.2	13 ± 1.9	1.3 ± 0.4

K562 cells were incubated with either control (K562) or producer cell supernatant (KRV) and cloned in soft agar in the presence or absence of geneticin G418.
Dishes labeled * contained too many colonies to count accurately.

favored form of DEAE–dextran has a molecular weight of 500 000 at a final concentration of 125 µg/ml. This method should be optimized for the cell type of interest, to check for the toxicity of the DEAE–dextran, and the incubation time of the cells adjusted accordingly. In addition, the duration and extent of the shock procedure should be investigated.

Another method using Polybrene which is claimed to be successful for both adherent [10] and suspension cells [11] is very similar to the DEAE–dextran method. It appears that the Polybrene binds both to the cell and the DNA; this method also utilizes a DMSO shock to facilitate the internalization of the foreign DNA.

A more recently introduced method is liposome-mediated gene transfer [12]. This involves the encapsulation of DNA in a lipid bilayer vesicle of L-α-phosphatidyl-L-serine, usually isolated from bovine brain. The phospholipid bilayer of the liposome is similar to that in the plasma membrane. When the DNA-loaded liposomes are added to the cells they are thought to fuse with the lipids in the plasma membrane; the DNA is subsequently released inside the cell and can then integrate into the genomic DNA. This is potentially an extremely powerful technique, but there are a number of parameters that have not yet been fully optimized, which can make it very unreliable. This could be due to batch problems in the purification of the phosphatidylserine, or some unknown and yet critical component in the preparation of the liposomes. There are a number of kits now on the market that use a technique based on liposome fusion. Information on these can be obtained from Gibco (Lipofectin™) or Promega (Transfectam™).

To transfect cells by protoplast fusion [13–15] the bacteria that are used to grow the plasmids are treated with lysozyme to digest the cell wall; the resulting protoplasts are intact bacteria that have only a bacterial plasma membrane remaining. The target eukaryotic cells are then fused with these protoplasts and their entire contents released into the cell.

Electroporation [16–18] is a method that may in time supersede many of the methods currently available for suspension cells. This involves applying a high-voltage pulse to the cells in the presence of plasmid DNA. The cell membrane is reversibly permeabilized and thus the DNA can enter the cells. This method has performed well on a number of different suspension cell types, although it does require specialist electroporation equipment.

Other methods of transfection include scrape loading [19], microinjection [20] and the use of single-stranded oligonucleotides [21–24].

Scrape loading has been used for adherent cells and involves mechanically wounding a culture in the presence of DNA. For reasons that are unclear, this appears to stimulate the cells to internalize the DNA associated with the membrane.

Microinjection uses extremely fine needles to introduce DNA directly into the nucleus of the target cell. The major limitation of this technique is the time required to introduce DNA into a sufficiently large number of cells when the injection must be done on an individual basis.

Single-stranded oligonucleotides cannot be used to generate stably transfected cells, but have been used successfully by a number of workers to introduce transiently antisense sequences into cells. The oligonucleotides are synthesized on a DNA synthesizer and are added directly to the cells in culture. It appears that short single-stranded DNA sequences can enter the cell with quite a high efficiency. It is unclear what maximum length of oligonucleotide can be used; also, little is known about the ability of double-stranded oligonucleotides to cross the cell membrane. A useful text that can be obtained free from Promega on request is *The Promega Protocols and Applications Guide* (1992). This benchtop guide contains a section on the transfection of cells and includes details of kits for transfection by calcium phosphate and DEAE–dextran, for both adherent and suspension cells in the Profection™ Mammalian Transfection systems section, along with a comprehensive list for further reading.

References

1. Graham, F.L. and Van Der Eb, A.J. (1973) *Virology*, **52**, 456.

2. Ford, T.C. and Graham, J.M. (1991) *An Introduction to Centrifugation*, BIOS Scientific Publishers, Oxford.

3. Sambrook, J., Fritsch, E.F. and Maniatis, T. (1989) *Molecular Cloning, a laboratory manual,* Cold Spring Harbor Laboratories, New York.

4. Davis, L.G., Dibner, M.D. and Batty J.F. (1986) *Basic Methods in Molecular Biology*, Elsevier.

5. Chu, G. and Sharp, P.A. (1981) *Gene*, **13**, 197.

6. Southern, P. and Berg, P. (1982) *J. Mol. Appl. Genet.*, **1**, 177.

7. Blochlinger, K. and Diggelmann, H. (1984) *Mol. Cell. Biol.*, **4**, 2929.

8. McCutchan, J.H. and Pagano, J.S. (1968) *J. Natl. Cancer Inst.*, **41**, 351.

9. Somapayrac, L.M. and Danna, K.J. (1981) *Proc. Natl. Acad. Sci. USA*, **78**, 7575.

10. Kawai, S. and Nishizawa, M. (1984) *Mol. Cell. Biol.*, **4**, 1172.

11. Chisholm, O. and Symonds, G. (1988) *Nucleic Acids Research*, **16**, 2352.

12. Itani, T., Ariga, H., Yamaguchi, N., Tadakuma, T. and Yasuda, T. (1987) *Gene*, **56**, 267.

13. Schaffner, W. (1980) *Proc. Natl. Acad. Sci. USA*, **77**, 2163.

14. Rassoulzadegan, M., Binetruy, B. and Cuzin, F. (1982) *Nature*, **295**, 257.

15. Yokoyama, K. and Imamoto, F. (1987) *Proc. Natl. Acad. Sci. USA*, **84**, 7363.

16. Potter, H.L., Weir, U. and Leder, P. (1984) *Proc. Natl. Acad. Sci. USA*, **81**, 7161.

17. Neumann, E., Schaefer-Ridder, M., Wang, Y. and Hofschneider, P.H. (1982). *EMBO J.*, **1**, 841.

18. Reisman, D. and Rotter, V. (1989) *Oncogene*, **4**, 945.

19. Fecheimer, M., Boylan, J.F., Parker, S., Sisken, J.E., Patel, G.L. and Zimmer, S.G. (1987) *Proc. Natl. Acad. Sci. USA*, **84**, 8463.

20. Capecchi, M.R. (1990) *Cell*, **22**, 479.

21. Cooney, M., Czernuszewicz, G., Postel, E.H., Flint, S.J. and Hogan, M.E. (1988) *Science*, **241**, 456.

22. Heikkila, R., Schwab, G., Wickstrom, E., Loke, S.L., Pluznik, D.H., Watt, R. and Neckers, L.M. (1987) *Nature*, **328**, 445.

23. Holt, J.T., Redner, R.L. and Nienhuis, A.W. (1988) *Mol. Cell. Biol.*, **8**, 963.

24. Wickstrom, E.L., Bacon T.A., Gonzalez, A., Freeman, D.L., Lyman, G.H. and Wickstrom, E. (1988) *Proc. Natl. Acad. Sci. USA*, **85**, 1028.

Appendix A

Glossary

Aliquot: a small sample.

Antigen: a molecule capable of inducing an immune response.

Aseptic: without microbial contamination.

Autologous: the same as, i.e. cells from the same person or strain of animals.

Buffer: a liquid which resists changes in pH.

Cell cycle: the process of cell division, comprising four parts (G_1,S,G_2,M) which describe DNA replication and cell division.

Cell culture: to grow cells *in vitro*.

Cell line: an established *in vitro* cell type (usually immortal or transformed).

Cloning: cell culture: selection of a single daughter cell from which to establish a culture of identical cells; molecular biology, manipulation of DNA replicas of DNA or RNA found in a tissue, organ or cell type.

Confluence: the measure of density of a cell culture.

Crisis (cell): the senescence or death of a cell culture (sometimes followed by an outgrowth of immortal cells).

Cryopreservation: to preserve by freezing at low temperature.

Cytotoxicity: to kill a cell (by another cell or drug).

Differentiation: a change in a cell causing it to lose or gain cell functions (usually non-reversible).

Fusogen: a substance which will cause membranes of adjacent cells to merge.

Genome: the DNA content of a cell.

Genotype: the gene content of a cell.

Hybrid: the result of two or more cells fusing to form a single entity.

Immortalization : changing a primary cell type with a limited life span *in vitro*, to a cell type with unlimited capacity to proliferate.

Immunization: to create an immune reaction by introducing a known antigen.

***In vitro*:** cell growth outside the body ('in glass').

***In vivo*:** cell growth within the body ('in life').

Karyotype: the chromosome complement of an individual cell.

Ligation: use of the enzyme ligase to join separate DNA molecules to form a single DNA molecule.

Medium: a buffered selection of components to support the growth of cells outside their normal environment.

Mitogen: stimulator of cell proliferation.

Monoclonal antibodies: antibodies derived from a single cell type producing a single antibody type.

Monolayer: growth of adherent cells on substratum.

Oncogene: gene implicated in the malignant transformation of cells, and thus also implicated in cancer.

Peripheral blood cells: those cells circulating in the bloodstream.

Phenotype: the expressed characteristics arising from the genotype.

Plasmid: loop of DNA capable of extrachromosomal replication in bacteria.

Polypropylene : a type of plastic which can withstand autoclaving.

Polystyrene : a type of plastic which cannot withstand autoclaving.

Proliferation : the division of cells producing further cells of the same type.

Quiescence: the resting state of a cell (G_0/G_1).

Retrovirus: a type of virus using RNA as its genetic material.

Senescence: cell or cells which have lost their growth potential.

Serial dilution: a series of repetitive dilutions.

Sterilize: remove microbial contaminants.

Subculturing: to perpetuate a cell line by reseeding a small number of cells into fresh medium.

Substratum: the surface upon which cells can rest or attach while growing.

Suspension culture: cells which grow without the need for adherence on to substratum.

Syngeneic: animals which have been inbred to produce a genetically similar/identical population.

Titre: concentration of infective agents.

Transfection: introduction of foreign DNA into a cell.

Transform: to transfect a bacterial cell, also to induce a cell to express abnormal characteristics (often similar to those found in tumor cells).

Trypsinization: to use the enzyme trypsin to remove adherence proteins from cell surfaces.

Appendix B

Suppliers

Applied Immune Sciences Inc. (AIS) (USA), 200 Constitution Drive, Menlo Park, CA 94025-1109, USA.

(UK distributors for AIS) Techgen International Ltd, Suite 8, 50 Sullivan Road, London SW6 3DX, UK.

American Type Culture Collection (ATCC), Sales and Marketing Department, 12301 Park Lawn Drive, Rockville, MD 20852, USA.

Becton-Dickinson (UK), Between Towns Road, Cowley, Oxford OX4 3LY, UK.

Becton-Dickinson (USA), Clay Adams Division, 299 Webro Road, Parsippany, NJ 07054, USA.

Boehringer Mannheim GmbH (UK), Boehringer Corporation (London) Ltd, Bell Lane, Lewes, Sussex BN7 1LG, UK.

Boehringer Mannheim GmbH, Biochemica, PO Box 31 01 20, D-6800 Mannheim 31, Germany.

Bibby-Sterilin, Tilling Drive, Stone, Staffs ST15 0SA, UK.

Cedarlane Laboratories Ltd, 5516-8th Line, R. R.2 Hornby, Ontario, Canada, L0P 1E0.

(UK distributors of Cedarlane) Vector Laboratories, 16 Wolfric Square, Bretton, Peterborough PE3 8RF, UK.

Dynal UK Ltd, Station House, 26 Grove Street, New Ferry, Wirral, Merseyside L62 5AZ, UK.

Dynal AS, PO Box 158, N-0212, Oslo 2, Norway.

European Collection of Animal Cell Cultures (ECACC), PHLS Centre for Applied Microbiology and Research, Porton Down, Salisbury SP4 0JG, UK.

Flow Laboratories Ltd (UK), Woodcock Hill, Harefield Road, Rickmansworth, Herts WD3 1PQ, UK.

Flow Laboratories Inc, PO Box 1065, Dublin, VA 24084, USA.

Gibco BRL Life Technologies Ltd (UK), Unit 4, Cowley Mill Trading Estate, Longbridge Way, Uxbridge, Middlesex UB8 2YG, UK.

Gibco BRL Life Technologies Inc (USA), 8400 Helgerman Court, Gaithersburg, MD 20877, USA.

ICN Biomedicals Ltd (European HQ), Eagle House, Peregrine Business Park, Gomm Road, High Wycombe, Bucks HP13 7DL, UK.

ICN Biomedicals Inc (USA), PO Box 19536, Irvine, CA 92713, USA.

Jencon (Scientific) Ltd, Cherrycourt Way, Industrial Estate, Stanbridge Road, Leighton Buzzard, Beds LU7 8UA, UK.

LEEC Ltd, Private Road No.7, Colwick Industrial Estate, Nottingham NG4 2AG, UK.

Luckham Ltd, Burgess Hill, Sussex RH15 9QN, UK.

Millipore UK Ltd, The Boulevard, Blackmoor Lane, Watford WD1 8YW, UK.

Millipore Corp (USA), 80 Ashby Road, Bedford, MA 01730, USA.

Nycomed UK Ltd, Nycomed House, 2111 Coventry Road, Sheldon, Birmingham B26 3EA, UK.

Nycomed AS, Pharma Diagnostica, Sandakervn 64, PO Box 4284 Torshov, N-0401, Oslo 4, Norway.

Payne Scientific, Hillside, Slough, Berks SL1 2RW, UK.

Promega Ltd (UK), Epsilon House, Enterprise Road, Chilworth Research Centre, Southampton SO1 7NS, UK.

Promega Corporation (USA), 2800 Woods Hollow Road, Madison, WI 53711-5399, USA.

Sartorius Ltd (UK), Longmead Business Centre, Blenheim Road, Epsom, Surrey KT19 9QN, UK.

Sartorius AG (HQ), PO Box 3243, 3400 Goettingen, Germany.

Sigma Chemical Company (UK), Fancy Road, Poole, Dorset BH17 7NH, UK.

Sigma Chemical Company (USA), 3Q50 Spruce Street, PO Box 14508, St Louis, MO 63178, USA.

TCS (Tissue Culture Services), 10 Henry Road, Slough, Berks, UK.

Travenol Laboratories Ltd, Caxton Way, Thetford, Norfolk, UK.

Travenol Laboratories Ltd, Deerfield, IL 60015, USA.

Universal Hospital Supplies Ltd, Wakefield, Yorkshire, UK.

Appendix C

Further Reading

Baserga, R. (ed) (1990), *Cell Growth and Division – a practical approach*, IRL Press, Oxford.

Cell Fusion (1984) Ciba Foundation Symposium 103, Pitman, London.

Feldman, M., Lamb, J.R., Woody, J.N. (eds) (1985) *Human T-Cell Clones*, Humana Press, New Jersey.

Freshney, R.I. (1987) *Culture of Animal Cells – a manual of basic technique*. Wiley-Liss, New York.

Lyderson, B.K. (ed) *Large-Scale Cell Culture Techniques*. Hanser Publishers.

Pollard, J.W., Walker, J.M. (1989) *Animal Cell Culture* Volume 5 of Methods in Molecular Biology. Humana Press, New Jersey.

Sikora, K., Smedley, H. (1984) *Monoclonal Antibodies*, Blackwell Scientific Publications, Oxford.

Index